DEAD MEN DO TELL TALES!

Have you ever wandered into a deserted house and *felt* the people who once lived there? Have you ever made a maiden voyage to a distant country and known the shock of recognition . . . the uneasy sensation that you had been there before? Have you ever awakened to see your dream become reality?

Feelings — premonitions — hunches — intuition. Common words for everyday occurences in the lives of most people. Yet if any of us asked where these sensations came from, we might be shocked to discover one of the most intriguing possibilities of all: that somehow, in some way, *we* may be able to transcend the prison of time. Here is a book that investigates that astounding possibility in all its forms.

TIME TRAVEL:

Myth or Reality?

RICHARD HEFFERN

PYRAMID BOOKS NEW YORK

TIME TRAVEL: MYTH OR REALITY?

A PYRAMID BOOK

Pyramid edition published April 1977

ISBN: 0-515-03719-2

Library of Congress Catalog Card Number: 76-57894

Printed in the United States of America

Pyramid Books are published by Pyramid Publications (Harcourt Brace Jovanovich, Inc.). Its trademarks, consisting of the word "Pyramid" and the portrayal of a pyramid, are registered in the United States Patent Office.

PYRAMID PUBLICATIONS
(Harcourt Brace Jovanovich, Inc.)
757 Third Avenue, New York, N.Y. 10017

To
the López family

Contents

List of Illustrations

Acknowledgments

The author wishes to express his appreciation and gratitude to the following authors, publishers, and publications for permission to include material from the following:

"The Tunbridge Wells Adventure" from *Man, Myth, and Magic* (vol. 35). Copyright 1971 by BPC Publishing Ltd.

"I Was Lost in the Fourth Dimension" by Miriam Golding, from *Fate* Magazine, September 1956. Copyright 1956 by Clark Publishing Co. Reprinted by special permission from *Fate* Magazine.

"Lost—Somewhere in Time" by Joseph E. Sainsbury, from *Fate* Magazine, June 1956. Copyright 1956 by Clark Publishing Co. Reprinted by special permission from *Fate* Magazine.

"Nostradamus" by Manly P. Hall. A lecture delivered at Town Hall, Curran Theatre, Los Angeles, Tuesday, January 25, 1944. Arthur Linkletter, Chairman and Moderator. Copyright 1944 by Manly Palmer Hall.

"Project Hermes" from *Two Inventors Return*. Published by the Borderland Sciences Research Foundation.

"Project Hermes Revived" from *Journal of Borderland Research* 21, No. 5 (July–August 1965), pp. 5–12.

The Statesman by Plato. Translated by J. B. Skemp, edited by Martin Ostwald. New York: Bobbs-Merrill Library of Liberal Arts. Copyright 1957 by The Liberal Arts Press. Reprinted by special permission.

Special thanks, also, to Lee Baxter, Rev. Roger De Board, Kenneth Eames, H. Murken, and Edward S. Schultz.

Introduction

Instinctively I feel that time, like sickness and suffering and evil, was not really meant to be. I do not think that time is an ultimate reality, but rather a part of the human predicament—a part of the confusion and chaos into which we humans are born. It is a phenomenon that we must ultimately grow beyond, at least in our awareness.

But in the normal course of events, time seems so much a part of everything that I do not even think to question it. I work and play and sleep in the matrix of a twenty-four hour day. I live surrounded by ample reminders of the unceasing passage of time. I grow older and those around me grow older. People form new friendships, change jobs, change plans, and move from one place to another. Houses, buildings, and roads appear where once there were orchards and vacant fields. My surroundings are in a constant flux of movement and change. Ideas, ideals, and modes of living seem to constantly change. I am reminded of Alvin Toffler's warning that the rapidly accelerating rate of change may be testing man's capacity for adaptation to its limit. Senseless, violent crimes, increasing occurrences of mental and emotional problems, and general widespread confusion all attest to the reality of what Toffler terms "future shock."

The relentless passage of time, as symbolized by the ticking of a clock and the removal of another page from the calendar on the wall seems, at times, to dominate every facet and aspect of my daily existence.

But it is not so. Contrary to appearances, time is not really omnipresent, even though I tend to forget this in the hectic routine of everyday living. On relatively infrequent occasions, I am able to catch a scene from the past—not an early memory, but something that took place in my surroundings, or something that may have involved a nearby object or person. Or I may be able to envision something from the future. A dream or a premonition that seems puzzling at the time may later come true.

In a recent instance, my wife informed me that she wanted to take a trip with her brother to attend her sister's wedding. Initially, I had no objections whatsoever, but just before she left, I had the feeling that something was wrong. Moments before she left, I thought of asking her not to go, but I said nothing. I dismissed my uneasiness. She told me exactly when she would return, and I know from past experience that she always calls me if a delay is anticipated. After they were a day late in returning, I became worried. When they still did not return, I became considerably alarmed. When I finally did hear from them, I learned that they had had serious troubles on the trip home, but was relieved to learn that they were not harmed.

Such an experience is merely a simple premonition; they are not at all uncommon. Nevertheless, they cannot be explained in the context of time.

If the reader can suppose, for the moment, that not all things can be explained in the context of time, that encounters with the past and future do occur, then time travel becomes at least a plausible concept, though it remains poorly understood. Fictional books on the subject abound, as the idea of time travel has long fascinated both writers and readers of science fiction. The reader may remember the following excerpt from one of the earliest and best-known science fiction classics, *The Time Machine* by H. G. Wells (1895):

> Now I want you to understand that this lever, being pressed over, sends the machine gliding into the future, and this other reverses the motion. This saddle represents the seat of the time traveller. Presently I am going to press this lever, and off the machine will go. It will vanish, pass into future time, and disappear. Have a good look at the thing. Look at the table, too, and satisfy yourselves there is no trickery. I don't want to waste this model, and then be told I'm a quack.

Unfortunately, time travel has traditionally been considered to be proper only in the realm of the science fiction writer and the imaginations of children. Aside from being an occasional topic of advanced theoretical physics, it has never been taken too seriously. Persons who have announced their involvement in the study of time travel have been looked upon with considerable skepticism at best, perhaps ridicule. A person who actually announces he is building a time machine might consider himself fortunate if he is not involuntarily confined.

It is a fact that certain people, at certain times and under certain conditions, have been able to break the time barrier and perceive things from the past and future. Psychometry, precognition, and retrospective clairvoyance are all examples of phenomena that occur outside of the normal matrix of time. These phenomena constitute one form of time travel in which only the mind penetrates the time barrier, while the physical body remains confined in the present. This type of experience, while in need of much additional investigation, is generally accepted as something that does occur.

The ultimate form of time travel, as visualized by H. G. Wells, has probably not yet been realized—and with good reason. The possibility of time travel is real, but the details remain poorly understood. I do not wish to imply that this situation is unusual. For centuries, birds and insects gave proof to man that flight was possible. But the secret of flight eluded even the genius of da Vinci. Today, no one denies the existence of light and gravity. But ask a scientist to explain exactly what light is, or what gravity is, and he cannot. The exact natures of these phenomena are yet unknown. Similarly, a thorough understanding of time travel has not yet been realized.

In this book, I have attempted to compile some of what is known about the subject of time travel. I have taken an anthological approach in order to present a multiplicity of different viewpoints and aspects of the subject. I wrote this book because I am fascinated with the concept of time travel. A breakthrough in man's understanding of time travel is forthcoming and imminent, and it will truly be one of mankind's greatest technological achievements.

PART I:
Lost in
Time and Space

1 Other Dimensions

Fictional characters in science fiction novels sometimes find themselves trapped in other dimensions. This is also true of fictional characters in science fiction television series such as *Twilight Zone, Star Trek,* and others. Many people have an intuitive idea of what is meant by the term, yet it is difficult to define.

One cannot actually travel to another dimension, because other dimensions are not separated from our own dimension by distance. Other dimensions, if they do exist, are all around us. They cannot be seen, heard, perceived, or measured unless one leaves this dimension and enters into another. In *Strange Mysteries of Time and Space* Harold T. Wilkins wrote:

> There is, so to speak, a "hole in the wall," or, to vary the metaphor, the occurrence of a vortex in matter-energy through which certain people of all social classes and ranks, animals, and objects, involuntarily and unexpectedly may pass, and from which there may, or may *not* be a return to the world from which they were "teleported."

Others have referred to a warp in the space-time continuum, which means essentially the same thing that Wilkins called a "vortex in matter-energy." In other words, on rare occasions, an opening forms in our dimension. The opening may close instantly, or it may stay open for a short time. These openings are the gateways to other dimensions, through which a person or other object may pass.

Upon passing through one of these openings, an object or person may (a) disappear forever, (b) return just as it entered, or (c) reappear in another location.

The first possibility will be dealt with first. The following section examines cases in which individuals apparently slipped into other dimensions, and did not return.

2 Mysterious Disappearances

Over the centuries, mankind has frequently been baffled by strange disappearances which seem to elude any rational explanation.

A famous case of recent times was that of Mr. David Lang. On September 23, 1880 he was on his Tennessee farm in the company of his wife and two children. He was about to leave on an errand, but turned around and walked back toward his house when he saw a friend, Judge August Peck, drive up in his horse and buggy. After taking just a few steps, David Lang suddenly vanished in front of everyone present. His family and Judge Peck watched in horrified amazement. Friends and neighbors searched the area well into the night, but they could not even find a weak spot in the ground; there was absolutely no rational explanation for his disappearance.

Mrs. Lang never lost hope that David might somehow return, but her hopes were in vain; David Lang was never seen again. His two children thought that they could hear his voice very faintly at the spot where their father had vanished. They heard him calling for help. Finally, the sound grew progressively fainter until it could no longer be heard.

The above example is only one of many similar happenings. In separate incidents, several children have mysteriously disappeared while stepping around a particular corner in the Devil's Gate area of California's

Angeles National Forest. In one case, several witnesses insisted that one of the boys vanished in plain view of everyone present. In each case, extensive searches of the area did not yield a single clue as to the cause of the disappearance.

Stories of a similar nature constituted one of the favorite topics of Ambrose Bierce. Bierce was once one of the better-known American writers, but today he is remembered primarily for *In the Midst of Life* and *The Devil's Dictionary*.

Bierce was born in Ohio in 1842, and was characterized by a certain morbid outlook that colored most of his writings. He wrote several accounts of unexplained, paranormal disappearances, most of which were published in *Can Such Things Be?* in a chapter entitled "Mysterious Disappearances." Bierce himself disappeared in a mysterious manner. His last written communication was a letter to his secretary dated December 16, 1913. It was mailed at Laredo, Texas, and contained numerous references to the Mexican Revolution, which was underway at that time. In the letter, Bierce stated that he was going to Mexico "with a pretty definite purpose which is not at present disclosable." Bierce was never seen nor heard from again. It is presumed that he died in late 1913, or sometime in 1914, although his body was never found.

The following short stories, alleged to be true accounts, are reprinted from *Can Such Things Be?*, first published in 1893.

THE SPOOK HOUSE

On the road leading north from Manchester, in eastern Kentucky, to Booneville, twenty miles away, stood, in 1862, a wooden plantation house of somewhat better quality than most of the dwellings in that region. The house was destroyed by fire in the year following—probably by some stragglers from the retreating column of General George W. Morgan, when he was driven from Cumberland Gap to the Ohio river by General Kirby Smith. At the time of its destruction, it had for four or five years been vacant. The fields about it were overgrown with brambles, the fences gone, even the few negro quarters, and outhouses generally, fallen partly into ruin by neglect and pillage; for the negroes and poor whites of the vicinity found in the building and fences an abundant supply of fuel, of which they availed themselves without hesitation, openly and by daylight. By daylight alone; after nightfall no human being except passing strangers ever went near the place.

It was known as the "Spook House." That it was tenanted by evil spirits, visible, audible and active, no one in all that region doubted any more than he doubted what he was told of Sundays by the traveling preacher. Its owner's opinion of the matter was unknown; he and his family had disappeared one night and no trace of them had ever been found. They left everything—household goods, clothing, provisions, the horses in the stable, the cows in the field, the negroes in the quarters—all as it stood; nothing was missing—except a man, a woman, three girls, a boy and a babe! It was not altogether surprising that a plantation where seven human beings could be simultaneously effaced and nobody the wiser should be under some suspicion.

One night in June, 1859, two citizens of Frankfort, Col. J. C. McArdle, a lawyer, and Judge Myron Veigh, of the State Militia, were driving from Booneville to Manchester. Their business was so important that they decided to push on, despite the darkness and the mutterings of an approaching storm, which eventually broke upon them just as they arrived opposite the "Spook House." The lightning was so incessant that they easily found their way through the gateway and into a shed, where they hitched and unharnessed their team. They then went to the house, through the rain, and knocked at all the doors without getting any response. Attributing this to the continuous uproar of the thunder they pushed at one of the doors, which yielded. They entered without further ceremony and closed the door. That instant they were in darkness and silence. Not a gleam of the lightning's unceasing blaze penetrated the windows or crevices; not a whisper of the awful tumult without reached them there. It was as if they had suddenly been stricken blind and deaf, and McArdle afterward said that for a moment he believed himself to have been killed by a stroke of lightning as he crossed the threshold. The rest of this adventure can as well be related in his own words, from the Frankfort *Advocate* of August 6, 1876:

"When I had somewhat recovered from the dazing effect of the transition from uproar to silence, my first impulse was to reopen the door which I had closed, and from the knob of which I was not conscious of having removed my hand; I felt it distinctly, still in the clasp of my fingers. My notion was to ascertain by stepping again into the storm whether I had been deprived of sight and hearing. I turned the door-knob and pulled upon the door. It led into another room!

"This apartment was suffused with a faint greenish light, the source of which I could not determine, making everything distinctly visible, though nothing was sharply defined. Everything, I say, but in truth the only objects within the blank stone walls of that room were human corpses. In number they were perhaps eight or ten—it may well be understood that I did not truly count them. They were of different ages, or rather sizes, from infancy up, and of both sexes. All were prostrate on the floor, excepting one, apparently a young woman, who sat up, her back supported by an angle of the wall. A babe was clasped in the arms of another and older woman. A half-grown lad lay face downward across the legs of a full-bearded man. One or two were nearly naked, and the hand of a young girl held the fragment of a gown which she had torn open at the breast. The bodies were in various stages of decay, all greatly shrunken in face and figure. Some were but little more than skeletons.

"While I stood stupefied with horror by this ghastly spectacle and still holding open the door, by some unaccountable perversity my attention was diverted from the shocking scene and concerned itself with trifles and details. Perhaps my mind, with an instinct of self-preservation, sought relief in matters which would relax its dangerous tension. Among other things, I observed that the door that I was holding open was of heavy iron plates, riveted. Equidistant from one another and from the top and bottom, three strong bolts protruded from the beveled edge. I turned the knob and they were retracted flush with the edge; released it, and they shot out. It was a spring lock. On the inside there was no knob, nor any kind of projection—a smooth surface of iron.

"While noting these things with an interest and attention which it now astonishes me to recall I felt myself thrust aside, and Judge Veigh, whom in the intensity and vicissitudes of my feelings I had altogether forgotten, pushed by me into the room. 'For God's sake,' I cried, 'do not go in there! Let us get out of this dreadful place!'

"He gave no heed to my entreaties, but (as fearless a gentleman as lived in all the South) walked quickly to the center of the room, knelt beside one of the bodies for a closer examination and tenderly raised its blackened and shriveled head in his hands. A strong disagreeable odor came through the doorway, completely overpowering me. My senses reeled; I felt myself falling, and in clutching at the edge of the door for support pushed it shut with a sharp click!

"I remember no more: six weeks later I recovered my reason in a hotel at Manchester, whither I had been taken by strangers the next day. For all these weeks I had suffered from a nervous fever, attended with constant delirium. I had been found lying in the road several miles away from the house; but how I had escaped from it to get there I never knew. On recovery, or as soon as my physicians permitted me to talk, I inquired the fate of Judge Veigh, whom (to quiet me, as I now know) they represented as well and at home.

"No one believed a word of my story, and who can

wonder? And who can imagine my grief when, arriving at my home in Frankfort two months later, I learned that Judge Veigh had never been heard of since that night? I then regretted bitterly the pride which since the first few days after the recovery of my reason had forbidden me to repeat my discredited story and insist upon its truth.

"With all that afterward occurred—the examination of the house; the failure to find any room corresponding to that which I have described; the attempt to have me adjudged insane, and my triumph over my accusers—the readers of the *Advocate* are familiar. After all these years I am still confident that excavations which I have neither the legal right to undertake nor the wealth to make would disclose the secret of the disappearance of my unhappy friend, and possibly of the former occupants and owners of the deserted and now destroyed house. I do not despair of yet bringing about such a search, and it is a source of deep grief to me that it has been delayed by the undeserved hostility and unwise incredulity of the family and friends of the late Judge Veigh."

Colonel McArdle died in Frankfort on the thirteenth day of December, in the year 1879.

After reading the preceding account, one might at first suppose that Colonel McArdle and Judge Veigh came upon a place where several people had met with death, either by accident or foul play. However, certain clues in the story indicate that the room into which McArdle and Veigh entered was actually of another dimension. Upon entering the house from a raging storm, "not a gleam of the lightning's unceasing blaze penetrated the windows or crevices; not a whisper of the awful tumult without reached them there." This led McArdle to seriously question whether he had been killed, or deprived of sight and hearing. If it had merely been a deserted house, lightning would have been clearly visible through the windows, and rain hitting the roof would have certainly been audible. Those who have entered other dimensions have often described them in similar terms; they are referred to as strangely quiet, lifeless places.

It is also significant that a thorough search of the house failed to reveal a room such as the one McArdle had described. Possibly the room belongs to another dimension which can only be entered at certain times, from which escape is nearly impossible. Perhaps Judge Veigh was the latest victim of a bizarre dimensional trap which had already claimed several others.

In the stories that follow, Ambrose Bierce describes cases in which the details of each victim's disappearance are unknown. All that is known is that the victim seemed to physically vanish, quite suddenly, from the face of the earth.

THE DIFFICULTY OF CROSSING A FIELD

One morning in July, 1854, a planter named Williamson, living six miles from Selma, Alabama, was sitting with his wife and a child on the veranda of his dwelling. Immediately in front of the house was a lawn, perhaps fifty yards in extent between the house and public road, or, as it was called, the "pike." Beyond this road lay a close-cropped pasture of some ten acres, level and without a tree, rock, or any natural or artificial object on its surface. At the time there was not even a domestic animal in the field. In another field, beyond the pasture, a dozen slaves were at work under an overseer.

Throwing away the stump of a cigar, the planter rose, saying: "I forgot to tell Andrew about those horses." Andrew was the overseer.

Williamson strolled leisurely down the gravel walk, plucking a flower as he went, passed across the road and into the pasture, pausing a moment as he closed the gate leading into it, to greet a passing neighbor, Armour Wren, who lived on an adjoining plantation. Mr. Wren was in an open carriage with his son James, a lad of thirteen. When he had driven some two hundred yards from the point of meeting, Mr. Wren said to his son: "I forgot to tell Mr. Williamson about those horses."

Mr. Wren had sold to Mr. Williamson some horses, which were to have been sent for that day, but for some reason not now remembered it would be inconvenient to deliver them until the morrow. The coachman was directed to drive back, and as the vehicle turned Williamson was seen by all three, walking leisurely across the pasture. At that moment one of the coach horses stumbled and came near falling. It had no more than fairly recovered itself when James Wren cried: "Why, father, what has become of Mr. Williamson?"

It is not the purpose of this narrative to answer that question.

Mr. Wren's strange account of the matter, given under oath in the course of legal proceedings relating to the Williamson estate, here follows:

"My son's exclamation caused me to look toward the spot where I had seen the deceased (*sic*) an instant before, but he was not there, nor was he anywhere visible. I cannot say that at the moment I was greatly startled, or realized the gravity of the occurrence, though I thought it singular. My son, however, was greatly astonished and kept repeating his question in different forms until we arrived at the gate. My black boy Sam was similarly affected, even in a greater degree, but I reckon more by my son's manner than by anything he had himself observed. (This sentence in the testimony was stricken out.) As we got out of the carriage at the gate of the field, and while Sam was hanging (*sic*) the team to the fence, Mrs. Williamson, with her child in her arms and followed by several servants, came running down the walk in great excitement, crying: 'He is gone, he is gone! O God! what an awful thing!' and many other such exclamations, which I do not distinctly recollect. I got from them the impression that they related to something more than the mere disappearance of her husband, even if that had occurred before her eyes. Her manner was wild, but not more so, I think, than was natural under the circumstances. I have no reason to think she had at that time lost her mind. I have never since seen nor heard of Mr. Williamson."

This testimony, as might have been expected, was corroborated in almost every particular by the only other eye-witness (if that is a proper term)—the lad James. Mrs. Williamson had lost her reason and the servants were, of course, not competent to testify. The boy James Wren had declared at first that he *saw* the disappearance, but there is nothing of this in his testimony given in court. None of the field hands working in the field to which Williamson was going had seen him at all, and the most rigorous search of the entire plantation and adjoining country failed to supply a clue. The most monstrous and grotesque fictions, originating with the blacks, were current in that part of the State for many years, and probably are to this day; but what has been here related is all that is certainly known of the matter. The courts decided that Williamson was dead, and his estate was distributed according to law.

AN UNFINISHED RACE

James Burne Worson was a shoemaker who lived in Leamington, Warwickshire, England. He had a little shop in one of the by-ways leading off the road to Warwick. In his humble sphere he was esteemed an honest man, although like many of his class in English towns he was somewhat addicted to drink. When in liquor he would make foolish wagers. On one of these too frequent occasions he was boasting of his prowess as a pedestrian and athlete, and the outcome was a match against nature. For a stake of one sovereign he undertook to run all the way to Coventry and back, a distance of something more than forty miles. This was on the 3rd day of September in 1873. He set out at once, the man with whom he had made the bet—whose name is not remembered—accompanied by Barham Wise, a linen draper, and Hamerson Burns, a photographer, I think, following in a light cart or wagon.

For several miles Worson went on very well, at an easy gait, without apparent fatigue, for he had really great powers of endurance and was not sufficiently intoxicated to enfeeble them. The three men in the wagon kept a short distance in the rear, giving him occasional friendly "chaff" or encouragement, as the spirit moved them. Suddenly—in the very middle of the roadway, not a dozen yards from them, and with their eyes full upon him—the man seemed to stumble, pitched headlong forward, uttered a terrible cry and vanished! He did not fall to the earth—he vanished before touching it. No trace of him was ever discovered.

After remaining at and about the spot for some time, with aimless irresolution, the three men returned to Leamington, told their astonishing story and were afterward taken into custody. But they were of good standing, had always been considered truthful, were sober at the time of the occurrence, and nothing ever transpired to discredit their sworn account of their extraordinary adventure, concerning the truth of which, nevertheless, public opinion was divided, throughout the United Kingdom. If they had something to conceal, their choice of means is certainly one of the most amazing ever made by sane human beings.

CHARLES ASHMORE'S TRAIL

The family of Christian Ashmore consisted of his wife, his mother, two grown daughters, and a son of sixteen years. They lived in Troy, New York, were well-to-do, respectable persons, and had many friends, some of whom, reading these lines, will doubtless learn for the first time the extraordinary fate of the young man. From Troy the Ashmores moved in 1871 or 1872 to Richmond, Indiana, and a year or two later to the vicinity of Quincy, Illinois, where Mr. Ashmore bought a farm and lived on it. At some little distance from the farmhouse was a spring with a constant flow of clear, cold water, whence the family derived its supply for domestic use at all seasons.

On the evening of the 9th of November in 1878,

at about nine o'clock, young Charles Ashmore left the family circle about the hearth, took a tin bucket and started toward the spring. As he did not return, the family became uneasy, and going to the door by which he had left the house, his father called without receiving an answer. He then lighted a lantern and with the eldest daughter, Martha, who insisted on accompanying him, went in search. A light snow had fallen, obliterating the path, but making the young man's trail conspicuous; each footprint was plainly defined. After going a little more than half-way—perhaps seventy-five yards—the father, who was in advance, halted, and elevating his lantern stood peering intently into the darkness ahead.

"What is the matter, father?" the girl asked.

This was the matter: the trail of the young man had abruptly ended, and all beyond was smooth, unbroken snow. The last footprints were as conspicuous as any in the line; the very nail-marks were distinctly visible. Mr. Ashmore looked upward, shading his eyes with his hat held between them and the lantern. The stars were shining; there was not a cloud in the sky; he was denied the explanation which had suggested itself, doubtful as it would have been—a new snowfall with a limit so plainly defined. Taking a wide circuit round the ultimate tracks, so as to leave them undisturbed for further examination, the man proceeded to the spring, the girl following, weak and terrified. Neither had spoken a word of what both had observed. The spring was covered with ice, hours old.

Returning to the house they noted the appearance of the snow on both sides of the trail its entire length. No tracks led away from it.

The morning light showed nothing more. Smooth, spotless, unbroken, the shallow snow lay everywhere.

Four days later the grief-stricken mother herself went to the spring for water. She came back and related that in passing the spot where the footprints had ended she had heard the voice of her son and had been eagerly calling to him, wandering about the place, as she had fancied the voice to be now in one direction, now in another, until she was exhausted with fatigue and emotion. Questioned as to what the voice had said, she was unable to tell, yet averred that the words were perfectly distinct. In a moment the entire family was at the place, but nothing was heard, and the voice was believed to be an hallucination caused by the mother's great anxiety and her disordered nerves. But for months afterward, at irregular intervals of a few days, the voice was heard by the several members of the family, and by others. All declared it unmistakably the voice of Charles Ashmore; all agreed that it seemed to come from a great distance, faintly, yet with entire distinctness of articulation; yet none could determine its direction, nor repeat its words. The intervals of silence grew longer and longer, the voice fainter and farther, and by midsummer it was heard no more.

If anybody knows the fate of Charles Ashmore it is probably his mother. She is dead.

It is interesting to note the similarities between the disappearance of David Lang and that of Charles Ashmore.

Both, apparently, could be faintly heard at the places where they had vanished. In both cases, their voices grew progressively fainter until they could no longer be heard.

Although it is only a matter of speculation, one might suppose that David Lang and Charles Ashmore continued to survive in the other dimensions into which they entered. Their eventual silence might have signified that their survival was brief. Or, perhaps, they tried unsuccessfully to reenter their own world at the places where they had vanished. Having failed to do so, perhaps they wandered off into other regions of their invisible habitations.

On the other hand, perhaps the interdimensional warp through which each man passed sealed itself gradually. Though the warp may have soon become too small to permit their escape, it may have been opened just far enough to permit the passage of faint sounds and noises. Then, after a while, the warp sealed itself completely.

All that is known with reasonable certainty is that these individuals did not merely become lost in the dimension in which they lived. Rather, they became lost in some invisible realm that is beyond human perception.

Ambrose Bierce concluded his book with these comments, which suggest a possible explanation for the mysterious disappearances he wrote about.

SCIENCE TO THE FRONT

In connection with this subject of "mysterious disappearance"—of which every memory is stored with abundant example—it is pertinent to note the belief of Dr. Hern, of Leipsic; not by way of explanation, unless the reader may choose to take it so, but because of its intrinsic interest as a singular speculation. This distinguished scientist has expounded his views in a book entitled "Verschwinden und Seine Theorie," which has attracted some attention, "particularly," says one writer, "among the followers of Hegel, and mathematicians who hold to the actual existence of a so-called non-Euclidean space—that is to say, of space which has more dimensions than length, breadth, and thickness—space in which it would be possible to tie a knot in an endless cord and to turn a rubber ball inside out without 'a solution of its continuity,' or in other words, without breaking or cracking it."

Dr. Hern believes that in the visible world there are void places—*vacua,* and something more—holes, as it were, through which animate and inanimate objects may fall into the invisible world and be seen and heard no more. The theory is something like this: Space is pervaded by luminiferous ether, which is a material thing—as much a substance as air or water, though almost infinitely more attenuated. All force, all forms of energy must be propagated in this; every process must take place in it which takes place at all. But let us suppose that cavities exist in this otherwise universal medium, as caverns exist in the earth, or cells in a Swiss cheese. In such a cavity there would be absolutely nothing. It would be such a vacuum as cannot be artificially produced; for if we pump the air from a receiver there remains the luminiferous ether. Through one of these cavities light could not pass, for there would be nothing to bear it. Sound could not come from it; nothing could be felt in it. It would not have a single one of the conditions necessary to the action of any of our senses. In such a void, in short, nothing whatever could occur. Now, in the words of the writer before quoted—the learned doctor himself nowhere puts it so concisely: "A man inclosed in such a closet could neither see nor be seen; neither hear nor be heard; neither feel nor be felt; neither live nor die, for both life and death are processes which can take place only where there is force, and in empty space no force could exist." Are these the awful conditions (some will ask) under which the friends of the lost are to think of them as existing, and doomed forever to exist?

Badly and imperfectly as here stated, Dr. Hern's theory, in so far as it professes to be an adequate explanation of "mysterious disappearances," is open to many obvious objections; to fewer as he states it himself in the "spacious volubility" of his book. But even as expounded by its author it does not explain, and in truth is incompatible with some incidents of, the occurrences related in these memoranda: for example, the sound of Charles Ashmore's voice. It is not my duty to indue facts and theories with affinity.

Dr. Hern is undoubtedly correct in affirming the existence of "holes, as it were, through which animate and inanimate objects may fall into the invisible world and be seen and heard no more."

However, he supposes that these holes are simply openings in the physical universe beyond which is utter nothingness. He states that "A man enclosed in such a closet could neither see nor be seen; neither hear nor be heard; neither feel nor be felt; neither live nor die . . ." He is at least partially correct. In a complete vacuum, sound cannot travel. Many physicists also believe that light cannot travel in utter nothingness. They believe that there is something in outer space—yet undetectable —that permits the passage of light. Thus, in utter nothingness a man could neither see nor be seen because light is necessary for the production of visual images.

Dr. Hern may be incorrect, however, in supposing that a man "in such a closet [could] neither live nor die." First of all, there would be nothing to support life—no food, water, or oxygen. Also, if a human being were suspended in a complete vacuum, his blood would probably boil due to the absence of atmospheric pressure. In short, a human being in this situation would probably die instantly.

According to Dr. Hern, then, there is only absolute nothingness beyond the visible universe. There is another theory, however, which suggests the existence of parallel universes as a possible alternate explanation.

3 Parallel Universes

Is it possible that other universes might exist beside our own universe, yet remain beyond the realm of normal human perception? Could it be that there are parallel universes that exist in a space and time totally different from our own?

This possibility has interested physicists and science fiction writers alike. In theory, it is possible. For instance, imagine an apartment building occupied by several separate families. If the building has good acoustical insulation, the families that share the building may be totally unaware of each other's activities. Similarly, our universe may actually be only a part of a larger universe. Though we may be unaware of it, we may be coexisting side by side with other universes, yet be unaware of their presence.

In theory, two parallel universes would not be separated by distance. Yet, they would exist in different space-time continuums. Our time would not correspond to their time. Our concept of space would be entirely different from their concept of space. For instance, a parallel universe would not exist in our past, present, or future; it would exist in a different kind of time altogether. Likewise, a parallel universe would neither be near, nor far away. It would exist in another kind of space.

A parallel universe would not necessarily conform to the laws that govern other universes. Natural laws that we take utterly for granted may have no validity in a parallel universe. For instance, in this world, it is easier to destroy than to create. The underlying principle behind this observation is that all physical objects tend toward maximum randomness. If I spill a jar of pennies, they will scatter in a random fashion on the floor; about half will land heads up, while the other half will land heads down. We have naturally learned to expect this because this phenomenon always occurs. Picture, for a moment, a world in which spilled pennies would land in neat, straight rows, all facing one direction. Although this would be an impossibility in our universe, it may be the *only* possibility in some other universe.

The most common argument against the theory of parallel universes poses the question: How can there be room for two universes to occupy the same place? The answer to this question lies in an analysis of what constitutes our physical universe. Essentially, our universe consists of atoms separated by vast distances. What appears to be a solid object is actually composed of *some* matter, but it is mostly space. A book, a rock, even a piece of lead is predominately made up of space. If it were possible to obtain a chunk of pure matter, a piece the size of a brick would weigh several tons!

Even the earth, which appears quite solid, is mostly space. Thus, there is certainly enough room for two universes to occupy the same place. There is, in fact, ample room for several universes. This would imply that what is generally termed "empty" space is not empty at all, but is occupied by things that the human being cannot perceive.

Another objection to the theory of parallel universes raises the question: If other universes exist all around us, is it likely that we would be totally oblivious to their existence? In answer to this, I ask the reader to consider the fact that there are many colors and sounds that the human senses cannot perceive. For example, the human ear cannot hear the high-pitched sound of a dog whistle, which dogs can hear at a great distance. Also, it has been discovered that bees can perceive infrared light, which is invisible to man.

Picture a person born with total colorblindness. It is impossible for him to even conceive of color. One could try for hours to describe red or blue or green to him, but he would be totally unable to picture any color. Yet, color would exist all around him; he would simply be unaware of its presence.

The parallel universe theory has been suggested as a possible explanation for flying saucer phenomena. Rather than originating from other planets, flying saucers may actually be capable of traveling back and forth between this universe and other, invisible universes.

4 Visits to Other Dimensions

Brief encounters with other dimensions of time and space have sometimes occurred without warning to individuals involved in the ordinary routine of daily living. Typically, these experiences happen to people who have no special interest in the occult, to say nothing of time travel. Also, within the stories to be described in this chapter there are several other common elements.

First, the individuals involved found that the other people they encountered seemed to be oblivious to their presence. This was not always the case; a notable exception is found in "Lost—Somewhere in Time" by Joseph Sainsbury. In that story, the subject appears to have communicated readily with those he encountered. Nevertheless, in most cases, the person who found himself or herself in another dimension was unable to communicate with other people in that dimension.

Secondly, the experiences were always referred to as being very real and lifelike. They did not seem hazy or ethereal in any way, and the experiences contained a great deal of intricate detail.

The possibility of hallucinations can safely be ruled out, because these were isolated experiences in the lives of sane, sober individuals. The subjects in question were not under undue stress or strain; they were merely going about their affairs in the usual way. Nor did these experiences have any of the usual morbidity characteristic of psychotic hallucinations; there were no elements of paranoia, guilt, or terror. Also, the elaborate patterns and images typically induced by hallucinogens were not present.

It would seem that these accounts represent actual, paranormal experiences that cannot be adequately explained on the basis of current scientific knowledge.

Lastly, there is one very important element common to all of the stories. It is so obvious, that it can easily be overlooked—each subject returned to tell about his experience. Unlike the cases recounted by Ambrose Bierce, the following subjects were fortunate enough to return.

Shopping in a small supermarket in 1968, Mrs. W. of Tunbridge Wells, England, seems to have stepped back into the past. There was nothing odd about the experience at the time until she discovered that the room she saw was not there, and had not been there for several years.

The Tunbridge Wells Adventure*

"Retrospective clairvoyance" is a term used for the type of psychic experience in which someone is apparently carried back into the past and sees events which occurred or places which existed years before.

Mrs. Charlotte W. is an old-age pensioner who lives with her husband near Tunbridge Wells in Kent, England. Mrs. W.—who before her experience had little or no knowledge of the supernatural—prefers to keep her identity secret to avoid curiosity seekers. She leads a fairly sheltered life. Once a week, she and her husband attend a whist drive** in their area, but apart

*"The Tunbridge Wells Adventure," from *Man, Myth, and Magic,* vol. 35. Copyright 1971 by BPC Publishing Ltd. Reprinted by special permission.

**A whist drive is when several groups of people get together in someone's house or hall to play whist. Whist is a card game played usually by four people, similar to bridge, in which points are scored.

from this and their Tuesday morning shopping trip into Tunbridge Wells, they rarely leave their cottage.

On Tuesday, 18 June 1968, Mr. and Mrs. W. set off into the town to buy their groceries. They left each other in the town centre to make their own purchases, arranging to meet later for coffee in the restaurant of a department store in High Street.

Mrs. W. had bought her usual weekly supplies but was anxious to obtain a tin of shortcake to give as a prize at her whist drive. Unable to find a suitable tin in the shops she normally used, she went into a small self-service store which was unfamiliar to her, and asked the assistant if the type of shortcake she required was available. It was not. Rather than leave empty-handed, Mrs. W. picked up two tins of soup and was looking around to see if any of the other goods on the shelves attracted her, when she noticed an opening in the left-hand wall of the store. It was the entrance, she saw, to a rectangular room measuring roughly twenty five feet by sixteen feet, and paneled in mahogany, in sharp contrast to the chrome and plastic decor of the store.

"The mahogany gave it a heavy appearance," said Mrs. W. "It had no windows that I could see, but was illuminated by a number of electric light bulbs with small frosted glass shades. In the room near the entrance, I saw two couples in mid-twentieth century dress, and can quite clearly remember the outfit of one of the women. It consisted of a beige felt hat tilted at an angle and trimmed on the left-hand side with a wisp of dark fur; her coat was also in beige, and was quite fashionable though rather long for 1968." Mrs. W. also saw about half a dozen men, dressed in dark lounge suits, sitting further back in the room.

"All these people were sitting at cream-topped tables drinking coffee and chatting, a fairly common scene in a small country town at eleven o'clock in the morning. There was a short counter near the left-hand wall and a glass-enclosed cash desk near the entrance, though I did not see a cashier."

It struck Mrs. W. as "rather odd" that she had not heard of this coffee shop in the town before, but she felt that it was an excellent idea from the point of view of the self-service store; the firm which owned it had long been established as tea and coffee importers.

"Another thing that struck me as odd was that I could not smell coffee or indeed any of the other odours familiar in a cafe, but in these days of air-conditioning I was not very surprised."

For a few moments, Mrs. W. debated with herself whether or not to try the coffee at her newly-discovered cafe, but decided against it and went out to join her husband at their normal rendezvous. She naturally told him about the "new coffee lounge" and they agreed to try it on the following Tuesday.

"A week later, after doing our shopping, we went back to the self-service store and headed for the left-hand wall where I had seen the cafe entrance," said Mrs. W. "There was no entrance; only a mammoth frozen foods cabinet. I felt rather shocked. My husband jokingly asked me what I had really been drinking the previous Tuesday. After recovering somewhat, I asked the senior assistant if the shop had a coffee lounge attached, but she shook her head and told me that I must have the wrong store. I left feeling an absolute fool."

After visiting her usual cafe, Mrs. W. recovered from her surprise sufficiently to persuade her husband to accompany her on an expedition in search of the mysterious coffee lounge. "After all," she said, "I knew what I had seen. I had described it to my husband the previous week. We visited the only two similar grocers' shops in the street. Neither had refreshment rooms attached to them. In any case, I knew one of them well, and I had never set foot in the other one before."

After thinking long and hard about the curious incident, Mrs. W. visited a Miss S., founder of a "psychic society" in Tunbridge Wells, and something of an expert in the supernatural. Unlike Mrs. W., Miss S. had lived in Tunbridge Wells for many years. Mrs. W. asked if she knew of any building similar to the one described. A small cinema, Mrs. W. remembered, had stood next to the self-service store until seven or eight years ago—did Miss S. remember any cafe being attached to it?

She did not. But after some thought she did remember that, at some time during the last war, she had visited the Tunbridge Wells Constitutional Club, which had been situated above and to the left of the modern store, and which had, as she remembered, mahogany paneling and small tables for refreshments.

"To cut a long story short," said Mrs. W., "I found the present address of the Constitutional Club, now on the outskirts of the town, and talked to the club steward on the phone. He had been with the club since 1919, with a short break during the Second World War. I tactfully asked him what the old club had been like, what sort of premises they were, and how many rooms there had been. He told me the club had been approached by a street door to the left of the self-service store I had visited, and a flight of stairs. On the second floor was an assembly room, and to the rear of that—to the left of the self-service store—was a small bar with tables for refreshments. Over this was a billiard room.

"He then told me that it was his job, when meetings were held, to set out the tables in the bar, and that coffee and soft drinks were served as well as alcoholic drinks.

"I asked him to describe the refreshments room and he did. His description tallied with everything I had seen. It was only then that I told him of my experience. He thought it quite remarkable."

As an interesting postscript, Mrs. W. discovered, twenty four hours after her visit to the self-service store on 18 June, that her last remaining aunt, aged eighty-eight, had died at about eleven o'clock in the morning of the 18th. [Mrs. W. said that her aunt] "had been interested in psychic matters all her life.

"It seemed a little odd," concluded Mrs. W., "but I can't really see the connection between her death and my seeing the coffee room. The whole business is most curious."

Dr. A. R. G. Owen of Trinity College, Cambridge, an expert on this type of phenomenon, comments: "Mrs. W. seems to be an honest and not very fanciful person, and there does not seem to be any obvious flaw in her story. Retrospective clairvoyance is a good deal rarer than ordinary contemporaneous clairvoyance, and there is no real explanation of it. In this case, if any connection between the aunt and the Constitutional Club or the shop, or its site, could be shown, it might provide a link."

"The Tunbridge Wells Adventure" is very valuable to the researcher of psychic phenomena for several reasons. First, the subject knew exactly when and where the experience occurred. Also, she had a good memory for details, and was able to give an excellent description of what she had seen. Lastly, it was possible to investigate what Mrs. W. had seen in the context of what had actually been there in the past. Thus, her experience appears to have been a genuine and convincing case of retrospective clairvoyance.

Unfortunately, not all such cases are as detailed and conclusive as this, and are therefore subject to further scrutiny, which is amply supplied by critics. There are many accounts that are indicative of retrospective clairvoyance, yet inconclusive. Such cases cannot be discounted, nor can they be proven. For instance, Edith Olivier, writing in *Without Knowing Mr. Walkley* (1938), tells of a drive she made from Devizes to Swindon, England, on a rainy October evening in 1916. She came upon what she believed was Avebury's Stone Age monument, composed of huge megaliths, which she had read about in a book. She stopped, got out of her car, and saw what appeared to be a number of quaint cottages built among the megaliths. She could hear the sounds of an English fair in progress, which the crowd seemed to be leisurely enjoying in spite of the pouring rain.

Miss Olivier later discovered that the traditional Avebury fair was discontinued in 1850, and that the megaliths were cleared away before 1800 to make use of the land for agricultural purposes. Although her experience does indicate that she may have witnessed a scene from the distant past, it could not be determined exactly where she had stopped. Thus, the incident remains somewhat inconclusive.

In the account that follows, Miriam Golding relates in considerable detail an experience she had upon stepping out of an elevator. The experience is unique insofar as she apparently met another individual who was as lost as she was in an unknown dimension.

I was Lost in the Fourth Dimension*

The recollection of an experience I had in the fall of 1934 still fills me with apprehension. Mentally, I always refer to this event as *the depot,* and I wonder what would have happened to me if, somehow, I had never come back.

I was a young girl. My husband was still my fiance and we lived in Chicago. We both were music students and had been to an afternoon concert. Finding we had ample time before keeping a dinner date with his family, we decided to browse around in a nearby music store. We took the elevator upwards and, once in the store, settled down on stools to study the latest scores and literature. I was paging through a magazine when my fiance, Stan, nudged me, pointing to the clock.

As far as I can be sure we *both* returned to the elevator but in the crush on the way down we got separated in the crowd. On arriving at what I thought was the main floor I tried to push my way out but was shoved back. The door closed again and we started down. I thought I could hear my fiance calling my name as the elevator descended below street level. Finally, there was that familiar thud which marked

*"I Was Lost in the Fourth Dimension" by Miriam Golding, from *Fate* Magazine, September 1956. Copyright 1956 by Clark Publishing Company. Reprinted by special permission from *Fate* Magazine.

the end of the line for old-fashioned elevators and the door opened once more. I was going to stay on to ride back up, but an unfriendly operator insisted "Everybody off!"

Leaving the car I was astonished to find myself in an immense place, a basement surely, but not of a downtown office building. There were boxes and crates stacked everywhere. Grimy, perspiring men were pushing carts or riding little trucks loaded with trunks and baggage. My eyes searched the place and I discovered, in one corner, a large iron staircase like a fire-escape. Walking toward it I thought I could see daylight above so I hurried to climb upward. Reaching the top, which was indeed above ground and in broad daylight, I was completely bewildered. There was no sign of the store I had left. Nothing else that should have been there was visible. There was nothing unusual in my surroundings but the place was totally unknown to me. I was in a large railroad station!

Crowds of travelers were hurrying about. There were the usual signs, *To Trains, Waiting Rooms, Lunch Counter, Tickets.* I was so engrossed with my surroundings that I almost knocked down one poor woman. I apologized, but she hadn't even noticed me. I could see no signs of arrivals or departures of trains, no timetables, and was curious (an understatement) to know where I was. Just then an announcer's voice rose above the din to read off a long string of names. However, I have seldom been able to understand railroad announcers and of this I caught not one word. Wandering around confusedly, I finally sighted the Information Booth. There was a long line and I joined it. Standing there I felt decidedly foolish about having to ask where I was, but when I reached the girl and voiced my question she seemed completely unaware of my existence. This was the last straw and I hurried away.

I followed the wall until I saw a *To The Street* sign and went outside into the open air. I still had no idea where I was. It was a beautiful day, warm, with a cloudless, blue sky that would have seemed like midsummer had not the leaves on the great trees along the avenue already turned gold, crimson, and orange.

There was a new red brick building going up across from the station. It looked like it might be a church. There were many people out on the street too, all looking healthy, pleasant, and contented. I smiled at some of the passers-by but received only blank stares. I heard friendly voices but couldn't understand a word. The place seemed so normal that I scarcely was frightened but who in such a situation could be other than confused and perplexed?

Wandering aimlessly along the street, I noticed ahead of me, a blond boy, probably in his teens, standing in the center of the walk staring in all directions. Nearing him, I stepped to one side to pass but as I did so he smiled eagerly, reached out to touch my arm as if to see if I was real. I stopped and smiled back.

He said haltingly, "I guess they—let you off at the wrong stop too?"

I immediately understood that, however fantastic, the same thing had happened to both of us. Our mutual plight created a bond and for want of anything better to do, we continued together down the broad avenue.

"This is weird," he remarked. "I was playing tennis back home and went to the locker-room to change my shoes. When I came out I was—in that depot."

"Where is home?" I asked.

"Why, Lincoln—Lincoln, Nebraska, of course." He sounded puzzled.

"But I started this—trip—in Chicago!" I told him.

We walked on, discussing everything we had ever heard or read of time-travel, teleportation, other space dimensions, but neither of us knew much of this sort of thing and we didn't get far.

After a while the street became less crowded. Ahead the road sloped downhill. Soon the town was behind us. We were in open country and ahead we saw the deep blue water of a lake or ocean. It was a welcome distraction and we raced down the hill, out onto the sandy beach where we flopped down on some large rocks to catch our breath. It was really lovely there, warm and fresh. The sun was dipping toward the water on the horizon now, so we assumed west was out that way. As we watched the sun drop we noticed a large sandbar not too far out. I thought I heard voices coming from there. Suddenly, I heard someone calling my name and, as I grew accustomed to looking into the bright sunlight, I saw with great surprise that one of the girls on the sandbar was my fiance's sister. There were others with her and they all waved and shouted again.

My new-found friend jumped up in excitement. "But this is wonderful!" he said. "Maybe they are some kind of—connection or—link." He searched for the right words and as he talked he scrambled out of all his clothes but his tennis shorts. "I'm going out there!" he exclaimed. "They *see* us! They *know* you! It's not far, I can make it in a few minutes!"

He dived into the waves and swam off. With an inner excitement I watched him go. He shouted back occasionally and swam on. The figures remained on the bar and their voices still reached me. But as he swam a queer thing happened; try as he might he could get no nearer to the sandbar. Then gradually the bar seemed to grow more distant. Finally he turned and swam back to shore where he dropped onto the sand in total discouragement. There seemed nothing to say; when we looked out again the bar had disappeared. There was no fog, no obscuring mist; the sun was quite low in the sky but it was still very light. However, the sandbar had vanished.

I can't imagine what we would have done next but suddenly I was enveloped in darkness. I felt as if I floated in space and then I was back on the stool in the music store! The magazine was still spread out before me. A clock was striking and the clerks were tidying up the counters preparatory to closing. I looked around for my fiance fully expecting to see him there, too. However, he was not in sight. I decided the best thing to do was to go straight to his home. This time I walked down the stairs!

When I got to my destination my fiance opened the door. He certainly looked relieved. He said he'd lost me on the elevator. After stepping out on the main floor he had been unable to locate me. Thinking I had gotten off on some other floor he had waited a while then finally decided to go home.

The rest of the family already was in the dining room and we followed them without further discussion. On entering I was more than a little surprised to see Stan's sister with the same friends I had seen on the sandbar.

She smiled as she said, "We saw you in town but you were so engrossed in each other you didn't even hear us!"

The following account was written by Joseph E. Sainsbury, shortly before his death in 1933. The Adventurer's Club of New York records these highlights of his career: Auditor, U.S. Marine Corps, 1897–1902; *U.S.S. Concord* during Battle of Manila Bay; U.S. Consular Service in China; Colonel of Engineers under Sun Yat-sen during the Chinese Republican Rebellion; U.S. Dept. of Justice, Bureau of Investigation during World War I.

After having stopped for a call at the Port of Cartagena, Colombia, it was time for him to return to his ship. Although he was gifted with an uncanny sense of direction, he could not find his way back. Had he lost his sense of direction—or had his surroundings changed?

Lost—Somewhere in Time*

While travelling in Central America, some years before World War I, the ship on which I was a passenger made a call at the Port of Cartagena, Colombia. We were to remain there for several hours. This gave me an opportunity to visit the city which had played such an important part in the history of Spanish America.

*"Lost—Somewhere in Time" by Joseph E. Sainsbury, from *Fate* Magazine, June 1956. Copyright 1956 by Clark Publishing Company. Reprinted by special permission from *Fate* Magazine.

My knowledge of Spanish is very meagre but I felt confident I could get by without an interpreter.

Upon leaving the ship I hailed a native carriage and travelled to the far end of the city. There I paid my cochero, dismissed him, and commenced to stroll around. After a while I started back toward my ship. I expected to find my way there without the least trouble as I was gifted with a keen sense of direction which always stood me in good stead during my travels all over the globe. So unerring was this sense that fellow travellers described it as uncanny.

Leisurely I started back to the port. Suddenly I realized that the sky had become overcast. Also I failed to note any familiar objects, objects seen while driving out. It became increasingly evident my vaunted sense of direction had failed me completely. I was lost! I hurried on for almost an hour, looking for some familiar point.

Becoming alarmed that I might be too late to catch my ship, and knowing that the skipper would not wait as we were already behind our schedule, I decided to get a guide. But not a vehicle was passing in the street. I met natives from time to time, and mustering up the little Spanish at my command, I approached them inquiring "¿Donde vapor?"

A stony stare and a "No comprendo, Señor." was the answer each time.

I began to feel strange, my sense of direction was lost—was I losing my individuality entirely? At this moment I began to think *intensely* of my boyhood dream of a medical career—why, I do not know. I tried to fight back these thoughts so foreign to my need, and to force my mind to devise a way to reach the port.

When I was about to despair, I suddenly saw a fine appearing, native gentleman. He approached, tipped his hat, and as he spoke I received the impression he believed me to be a physician. I tried to protest, but he could not, or would not understand me.

Now another queer thing happened. I found myself speaking Spanish as fluently as he. After explaining my predicament to him, he informed me he was about to make a short call at a house, and if I would accompany him he would be pleased to see me aboard my ship in plenty of time. I agreed unhesitatingly and we walked on together.

We arrived at a fine residence, a wide, low house, built around a court or enclosure which was open to the sky. My companion invited me into the large, well-furnished hallway, with its couches, chairs, and mirrors. At the far end, two long French windows opened into the patio where flowers and attractive walks were visible.

Here my guide excused himself for a minute and went into an adjoining room. I heard a number of voices, then he re-appeared at the door and invited me in.

Entering, I saw about twenty-five people standing around a large bed, which occupied the center of a very large room. On it a young woman lay. She was entirely covered with a sheet, except her head, which was emaciated. Her cheeks were hollow, her eyes sunken into her head. My conductor introduced me as a distinguished medico Americano, and strangely, this time I did not correct him.

Turning to one of the señoras he said further that he had given her daughter up to die, but that the Buen Dios had sent me in answer to her prayers. She pulled down the sheet, exposing the naked body of the most emaciated being I had ever seen, not even barring the wretched famine sufferers of India and China. The flesh around the girl's ribs, pelvis, knees and arms had shrunken so that the bones stood out prominently. She was a living skeleton.

Quickly recovering her body, I ordered all out of the room, with the exception of the girl's mother, who brought me a chair. Then taking a notebook from my pocket, I tore out four pages, wrote rapidly two prescriptions and made one copy of each. The copies I kept and the prescriptions I ordered filled immediately.

I have never understood how all this came about for I was not qualified, in any way, to pose as a physician. Something beyond my control was directing me.

I gave instructions regarding the case, in rapid Spanish, advising the mother to follow them if she would have the patient recover. Then stepping to the bedside, where the girl lay staring at me like one fascinated, I told her she must have implicit faith in the good Lord, as he had willed her recovery. I said adios, and departed with the young man as planned.

I noticed, after leaving the sickroom, that unaccountably my flow of Spanish left me. It was with difficulty that I caught the drift of my young friend's remarks. At the same time my sense of direction came back to me and I was able to locate many convents and other points of interest I had seen. I could have found my way unaided now but the young man at my side came the entire distance with me, finally bidding me adios with tears in his eyes.

After two days at sea I became worried over the writing of those prescriptions, but hiding my anxiety I casually showed them to the ship's surgeon, asking his opinion regarding their efficacy in the treatment of the disease for which they were intended. He was impressed with the perfect Latin in which they were couched, and said that they were the best known to the medical profession for the treatment of the disease I mentioned.

A few months later I returned to Cartagena on the same ship. Upon my arrival, my young Colombian friend was on deck. I had forgotten to give him my name but he had described me to the Captain, who sent for me.

Together my friend and I went to the home of my former patient. I met one whom I could hardly believe to be the same girl. While she was still thin, she was gaining rapidly. The emaciated look was gone, and before me was a beautiful woman who soon was to become the bride of my conductor.

A similar story was recounted in the April 1959 issue of *Fate*. It describes the experience of Mr. and Mrs. Peterson of Keokuk, Iowa, and their four children. They were on their way home after a weekend visit to Missouri in 1935. They took a shortcut off the main road, which led them to the rim of an unfamiliar valley. In the valley, they saw what appeared to be a quaint rural community of another time period. The buildings, the clothing styles, the primitive tools and implements, were all suggestive of a previous century.

They inquired of people familiar with the area, but no one had ever heard of such a settlement. Nor were the Petersons ever able to find it again.

Did the Petersons somehow travel back through time? If so, they, too, were fortunate enough to return to the present.

In each of the preceding accounts, there is one very important bit of information that is missing—did the people that had these experiences remain physically present in this dimension? In other words, if an observer had watched Mrs. W. when she saw the coffee shop, what would he have seen? Would Mrs. W. have simply disappeared right before his eyes, or would she have been visible to an observer? Likewise, would Miriam Golding and Joseph E. Sainsbury have been visible to an outside observer? Did the Petersons slip into some sort of time warp, or could they have been seen by other motorists while they witnessed the quaint rural settlement?

Although there is no way to answer these questions with certainty, another account which appeared in *Fate* does offer a clue. In a story submitted by J. P. J. Chapman of Poole, Dorset, England, he and his wife went in search of some wildflowers for a forthcoming charity event. They decided to look in a vacant area near some apartment buildings that contained an abundance of wild growth.

Mr. Chapman walked a short distance to some primroses. When he glanced up again, the apartment buildings in the background had disappeared. He stated that a "vast open space of nothingness surrounded me." Mr. Chapman apparently had some previous knowledge of other dimensions, and was clever enough to mark the ground where he entered this strange region.

He walked in the direction where the apartment buildings had been, but encountered nothing but a vast, empty space. The sun was in the sky, and the ground was below his feet, but he saw no signs of people, traffic, or buildings of any kind.

Thinking that his wife might be concerned about his absence, he turned and walked back toward the mark he had made on the ground. After passing the mark, he found that everything was suddenly normal again.

His wife had been worried about him. She stated that she had looked everywhere but could not see him.

It would seem that Mr. Chapman physically left this dimension, and physically entered another one. He did something that very, very few persons would have done —he marked the spot where he entered. Perhaps for this reason alone, he was able to return to his own dimension.

5 Mysterious Reappearances

There are also recorded cases in which individuals have apparently slipped into interdimensional warps, from which they reappear in entirely different locations.

Charles Fort, who was a noted investigator of paranormal phenomena, suspected that this does occur. He cited the example of the strange appearance of a man in High Street, Chatham, England, on the evening of January 6, 1914. What was so strange about this appearance is that the man seemingly appeared out of nowhere, *absolutely naked,* on a very cold evening. The man "ran up and down the street until a policeman caught him." Unable to account for himself, he was diagnosed as "insane" by the doctors.

Writing in *Lo!,* Fort continues:

> This naked man of Chatham appeared suddenly. Nobody had seen him on his way to his appearing-point. His clothes were searched for, but could not be found. Nowhere near Chatham was anybody reported missing.

Fort suggested that these unexplained appearances might occur when someone steps into another dimension in one part of the world, and suddenly reappears in another part of the world. Nevertheless, he stated that he had never "satisfactorily come upon [a case] in which somebody has stated that he was walking, say, in a street in New York, and was suddenly seized upon and set down somewhere, say in Siberia, or Romford." He concluded:

> If human beings ever have been teleported, and, if some mysterious appearances of human beings be considered otherwise unaccountable, an effect of the experience is effacement of memory [amnesia].

Fort died in 1932 and did not live long enough to hear of an account that appeared some years ago in a Pennsylvania newspaper (Lock Haven *Express,* June 7, 1956). It told the story of Thomas Kessell who inexplicably found himself on a New York City street, with no recollection of how he had arrived there. He was employed in a brewery in South Africa, and his last memory was that of leaving a Johannesburg tavern in April.

Kessell apparently experienced amnesia for the period between his disappearance in April and his reappearance in June, but he was fully aware of where he had come from and where he had arrived.

Even more recently, on June 3, 1968, Dr. and Mrs. Gerardo Vidal were returning from a family reunion at a residence just outside of Buenos Aires, Argentina. On their way home, they were suddenly enveloped in what they described as a dense fog bank. Later, they found themselves on an unfamiliar highway in unfamiliar surroundings. After asking several people where they were, they soon discovered that they had been mysteriously transported to a road outside of Mexico City.

Dr. Vidal phoned his relatives from Mexico City, who had become quite worried. The phone call was made two days after the mysterious disappearance had occured; Dr. and Mrs. Vidal had no recollection of what happened after their car was enveloped in fog. They returned home aboard a plane nonstop from Mexico to Argentina.

Several similar incidents were subsequently brought to light in which certain persons suddenly found themselves transported to Mexico from some part of South America. It is as though some sort of interdimensional tunnel connects these two parts of the world. It occasionally opens up, allowing these strange transports to occur.

6 Time Anomalies

Since 1840, numerous strange incidents have occured in a northwestern area of the Atlantic known as the Bermuda Triangle. Planes, ships, and entire crews have mysteriously disappeared for no apparent cause.

I will not attempt to reiterate the many, well-documented cases about which much has been written. Rather, I would like to focus on a few isolated cases in which the victims (for lack of a better word) left clues before they vanished, concerning what they were about to encounter, and those cases in which individuals experienced unexplained phenomena in the area of the Bermuda Triangle, but arrived unharmed at their destinations.

One of the most revealing disappearances occurred in 1945 to five TBM Avenger torpedo bombers which had left Fort Lauderdale Naval Air Station for routine maneuvers. A short while later, all five planes disappeared and were never seen or heard from again. The planes had left the ground at 2:00 P.M. on December 5, 1945. The following account is taken from *Invisible Residents* by Ivan T. Sanderson:

> The first radioed communication from the lead plane did not come in to the base until 3:35 P.M.—and, curiously, they did not request landing instructions but instead reported: "Calling tower this is an emergency . . . We seem to be off course . . . We cannot see land . . . (Repeat) . . . We cannot see land." Asked by the tower for their position, the astonishing reply came back: "We're not sure of our position. We cannot be sure just where we are. We seem to be lost."

> Advised by the tower to head due west, the even more astonishing reply came back: "We don't know which way is west. Everything is wrong . . . strange. We cannot be sure of any direction. Even the ocean doesn't look as it should."

The flight leader apparently panicked and turned over the flight command to another pilot, who seemed equally disoriented in his reports to the control tower. A few words from the new flight commander were heard, then silence.

All the pilots were experienced, and weather conditions were excellent. The peculiar disorientation that seemed to prevail during the incident would seem to suggest that the five planes were on the threshold of another dimension. The flight leaders did not report any tangible obstacles that might interfere with the completion of the flight. They did report, however, that nothing seemed as it should be. Perhaps they entered another dimension from which they could never return. An extensive search was made of the area, but it revealed no trace of the missing planes.

Sanderson also mentioned two letters he received, each from a military pilot. The first pilot recounts an incident that occurred while he was heading south over Korea in a B-26 in the fall or winter of 1955. His altitude was 7,000 feet, and the B-26 was indicating 285 knots, which is close to the maximum speed for that model. Yet his ground speed was 550 knots, indicating he had to be traveling in a wind of 265 knots. A wind of 265 knots at 7,000 feet in a southerly direction is unheard of, and regarded as a near impossibility. At any rate, it would certainly be noticeable on the ground. below, and no unusual conditions were observed on the ground. Also, the plane would have experienced severe turbulence while entering and leaving its southerly flight path, which it did not. Although the wind was actually forecast to be 20—30 knots, the plane *seemed* to be reacting to a wind ten times as great.

The second letter recounted a similar incident in which an Air Force pilot was en route from Kwajalein Island to Guam aboard an Air Guard C-97 at 12,000 feet. During his flight the ground speed suddenly increased as though he was being pushed by a 200 knot wind. The ground speed then returned to normal as though the 200 knot wind had left as mysteriously as it had appeared. In any event, the C-97 had covered 340 nautical miles in one hour of time, which was unheard of for that model at that altitude.

Sanderson had the two stories analyzed in detail by a high-ranking airline pilot whose only explanation was that these incredible winds must have suddenly arisen in very narrow, confined regions, and died down suddenly but gently, so that no turburlence was felt. "Something very weird is going on here" was the final comment.

Rather than accepting this explanation, Sanderson proposes a very interesting theory. He suggests that the planes may have become trapped in local "time anomalies" in which time ran slower. Thus, he states, "the planes would have comparatively longer to get farther," and arrive early. He surmises that if a plane entered a time anomaly in which time ran faster, the plane would be late, and the pilot would undoubtedly attribute it to head winds.

Time anomalies may not explain all of the disappearances that have occurred in the Bermuda Triangle, but the above examples do suggest the possibility that warps in time and space may occasionally open up in that region. In some cases, vessels may have been trapped in these warps forever. Indeed this may have been the fate of the five TBM Avenger torpedo bombers. On the other hand, the two military pilots were apparently able to emerge from these strange regions unharmed.

In addition to strange disappearances of human beings, there have also been reported cases of objects falling into invisible holes from which they are never recovered. Sometimes, the object in question may reappear later in another location. Harold T. Wilkins, writing in *Strange Mysteries of Time and Space,* describes an incident in which a man at sea accidentally dropped his knife overboard. At that same moment, his wife (who was at home) was startled to see the same knife fall from the ceiling in their kitchen and stick in the table.

Just as objects fall into these interdimensional holes, objects seem to fall out of the holes as well. Almost every imaginable object has been reported, green rain, pieces of red meat, live fish, cookies—even alligators. A peculiar substance termed "angel hair" is frequently reported following sightings of UFOs. It is a fine, white stringy material that falls from the sky in areas where flying saucers have been witnessed. These objects often fall out of clear, cloudless skies, void of even an airplane that could be the instrument of a practical joke.

Equally mysterious are several actual locations on earth which seem to belong to another world. These places seem to almost be exempt from the laws of nature.

One such place is Magnetic Hill near Moncton in New Brunswick, Canada. Automobiles, rubber balls, even water travels uphill with ease at this strange spot. The force is not magnetic because non-metal objects are affected also. At Magnetic Hill, gravity seems to work in reverse.

Another strange place in which things are not as they should be is the Oregon vortex along Sardine

Creek near Grant's pass, Oregon. The Oregon vortex is a circular region, about 165 feet in diameter. A strange force pulls people and other objects toward the center of the vortex, so that it is necessary to lean away from the center in order to maintain one's balance. Hanging objects lean toward the center. Objects will even roll uphill on an inclined surface toward the center of the vortex.

Scientific instruments confirm the presence of the force, but scientists are at a loss to explain it.

The strange forces at work at Magnetic Hill and at the Oregon vortex can be witnessed by anyone. However, there may be some equally strange spots on the earth that can be perceived only by certain susceptible individuals. For example, Brad Steiger, in *Strange Disappearances,* describes a man with the uncanny ability to pass through doors leading into other dimensions. He says that some of these doors lead into eerie, lifeless places, devoid of sound or motion. Others, he says, lead into the past or into the future of this world.

If such holes in time and space do, in fact, exist, man cannot be content to watch objects fall through them from time to time. Hopefully, our knowledge will advance to the point where it will be possible to understand how these phenomena occur.

PART II: Glimpses of the Past and the Future

7 Clairvoyant Dreams

In Genesis the biblical story is told of how the Egyptian Pharaoh had two perplexing dreams, similar in their content. First, he dreamt that seven ugly, thin cows devoured seven fat, healthy cows that were grazing at the banks of the Nile River. In the second dream, seven thin ears of grain devoured seven fat, full ears.

Pharaoh's magicians and wise men were unable to interpret the dream for him. Finally, he summoned Joseph, who told Pharaoh the meaning of his dreams. Joseph explained that there would be seven years of plenty in Egypt, followed by seven years of famine. He advised Pharaoh to store up food during the years of plenty to prepare for the famine. This occurred just as Joseph had predicted.

Although this is perhaps the earliest recorded account of a prophetic dream, it is by no means the only such account. History is filled with many incidents involving prophetic dreams. Many public figures have had dreams that foretold events in their lives, sometimes their own deaths.

A classic treatise on the prophetic nature of certain dreams was written by an English researcher, J. W. Dunne, who wrote *An Experiment with Time* (1927). Based on numerous case histories, the book concludes that many dreams do, in fact, deal with events of the past, as well as events that occur in the future, after the dream has taken place.

Dunne himself often had clairvoyant dreams. In the

beginning of the book, he recounts a dream that first drew his attention to this unusual faculty. Dunne had dreamt that his watch had stopped at precisely half past four. He then woke up, lit a match, and went to see if the watch had really stopped. He found the watch lying on his chest of drawers, and it had stopped at precisely half past four.

Dunne reasoned that it had probably stopped the day before, and that he had probably forgotten about it until he had the dream. He rewound the watch, but he did not know what time it was; consequently, he did not bother to reset it.

The next morning, he "made straight for the nearest clock, with the object of setting the watch right." To his surprise, the watch was off by two or three minutes. Dunne was forced to conclude that the watch must have stopped at four-thirty—at the exact time of the dream. He reasoned that the two or three minutes could be accounted for by the time it took him to wake up, light the match, and wind the watch.

Dunne had another dream in the autumn of 1913 in which he saw a northbound train that had fallen over the embankment at a place he recognized as just north of the Firth of Forth Bridge in Scotland. He felt that the incident would occur sometime in the following spring. On April 14, 1914 a mail train called the *Flying Scotsman* plunged down the embankment about fifteen miles north of the Forth Bridge, shortly after crossing it. His sister was able to confirm that he mentioned the dream to her the morning after he had it, thus ruling out any possibility that Dunne might have imagined it.

In another dream, Dunne dreamt that he was walking down a pathway between two fields. A tall fence consisting of iron railings separated him from a horse that was behaving in a wild and frenzied manner. A moment later, he heard the sound of hoofs behind him, and ran as fast as he could from the horse, which had somehow managed to escape.

The next day, Dunne (in real life) went fishing with his brother, and experienced an eerie sensation when he saw a horse in a nearby field behaving just as the one he had dreamt about. There were certain minor details that were different. Rather than iron railings, there was a wooden fence. Also, the horse in the dream was much larger than the one in real life.

At any rate, Dunne concluded that the wooden fence was substantial enough to keep the horse contained, and resumed his fishing. Suddenly, his brother shouted, "Look out!" The horse had escaped his confinement, charged past Dunne and his brother, and landed in the river. The horse emerged somewhat sobered by the experience, and no longer acted in a threatening manner.

Dunne had developed an interesting technique in order to remember his dreams. He kept a pencil and paper at the side of his bed, and wrote down his dreams immediately after waking from them, no matter how insignificant a dream might seem at the time. Later dream researchers have regarded this technique as indispensable because of the fleeting quality characteristic of dreams; nearly all dreams are forgotten moments after they occur. Many people who insist that they do not dream at all simply forget their dreams. Everyone, in fact, dreams many times each night but all, or nearly all, are forgotten.

Dunne surmised that most people probably have clairvoyant dreams, but simply do not bother to write them down; consequently, they are quickly forgotten. To test his hypothesis, he persuaded several friends and relatives to keep records of their dreams.

The experiments were not purely scientific, and there was certainly a considerable margin of error. However, with these factors taken into account, Dunne concluded that precognitive dreams are quite common. In fact, he stated that dreams probably contain as many references to the future as references to the past.

Dunne devoted a large part of his book to a theoretical analysis of the nature of time based on his findings. His conclusions are exceedingly complex, and resort to complex mathematical formulas that are beyond the scope of this book.

8 Washington's Vision at Valley Forge

It is said that George Washington had a prophetic vision at Valley Forge in which the future of the United States was revealed to him in symbolic fashion. In a very old account of Washington's vision, Wesley Bradshaw relates the information from Anthony Sherman, who had been a soldier in the Revolutionary Army. Sherman tells how he had seen Washington emerge from his quarters one day looking unusually pale. He quotes the words that Washington spoke:

> I do not know whether it was owing to the anxiety of my mind, or what, but this afternoon, as I was sitting at this very table engaged in preparing a dispatch, something in the apartment seemed to disturb me. Looking up, I beheld standing exactly opposite me a singularly beautiful female. So astonished was I, for I had given strict orders not to be disturbed, that it was some moments before I found language to inquire the cause of her presence.
>
> A second, third and even fourth time did I repeat the question, but received no answer from my mysterious visitor other than a slight raising of her eyes. But this time I felt a strange sensation spreading through me. I would have risen, but the riveted gaze of the vision before me rendered volition impossible. I essayed once more to address her, but my tongue had become paralyzed. A new influence, mysterious, potent, irresistible, took possession of me. All I could do was to gaze steadily, vacantly, at my unknown visi-

tant. Gradually the surrounding atmosphere seemed as though becoming filled with sensations and grew luminous. Everything about me seemed to rarify, the mysterious visitor herself becoming more airy, and yet, even more distinct to my sight than before. Now I began to feel as one dying, or rather to experience the sensations which I have sometimes imagined accompany dissolution. I did not think, I did not reason, I did not move; all were alike impossible. I was only conscious of gazing fixedly, vacantly, at my companion.

Washington then began to have a detailed vision veiled in symbolism. Nevertheless, the symbolism is not nearly as cryptic as that contained in the Apocalypse (Book of Revelation) of the Bible. The vision entailed three major crises:

> Three perils will come upon the republic. The most fearful is the second, passing which, the whole world united shall never be able to prevail against her.

The first peril may have been the Revolutionary War, and is described in this way:

> At that moment I beheld a dark, shadowy being, like an angel standing or rather floating in mid-air, between Europe and America. Dipping water out of the ocean in the hollow of each hand, he sprinkled some water upon America with his right hand, while he cast upon Europe some with his left. Immediately a dark cloud arose from each of those countries, and joined in mid-ocean. For a while it remained stationary, and then moved slowly westward, until it enveloped America in its murky folds. Sharp flashes of lightning now gleamed throughout it at intervals, and I heard the smothered groans and cries of the American people.

The second peril clearly refers to the Civil War:

> I saw an ill-omened spectre approaching our land. It flitted slowly and heavily over every village, town, and city of the latter, the inhabitants of which presently set themselves in battle array, one against the other. As I continued looking, I saw a bright angel, on whose brow rested a crown of light, on which was traced the word "Union," bearing the American flag, which he placed between the divided nations, and said, "Remember, ye are brethren."
>
> Instantly the inhabitants, casting from them their weapons, became friends once more, and united around the national standard.

The third peril described in the vision is one that apparently has not yet taken place, if, indeed, it ever will. It describes a situation in which the United States is invaded by foreign troops; the war is apparently fought on American soil:

> Then my eyes looked upon a fearful scene. From each of these countries [Europe, Asia, and Africa] arose, thick black clouds, which soon joined into one, and throughout the mass gleamed a dark-red light, by which I saw hordes of armed men, who moving with the cloud, marched by land and sailed by sea to America, which country was presently enveloped in the column of the cloud. And I dimly saw these vast armies devastate the whole country and pillage and burn villages, cities and towns that I had beheld springing up.

Yet in spite of the grim picture, the vision ended with words of optimism and encouragement:

> While the stars remain and the heavens send down dews upon the earth, so long shall the republic last.

After the vision, Washington came to this conclusion concerning the significance of what he had witnessed:

> With these words the figure vanished. I started to my feet, and felt that I had been shown the birth, progress, and destiny of the United States. In union she will have strength, in disunion her destruction.

9 Psychometry

Psychometry is a term frequently encountered in the study of paranormal phenomena. It is based on the theory that a person imparts an enduring record of his thoughts, actions, feelings, experiences, and emotions to everything he comes in contact with, provided there is some degree of emotional involvement between the person and that which he comes in contact with.

The object or place is believed to record this information just as a videotape recorder records both images and sounds. If one did not have a device that could play back the videotape, he would be at a loss to detect any recorded information at all. Similarly, an object or place gives no superficial clues as to the people and events that have touched it.

Psychometry is an extrasensory means of playing back the recorded information. A psychometer (person who possesses psychometric abilities) could, for instance, enter a room where a murder had taken place. Depending on his degree of sensitivity, he could perceive the people, objects, and emotions that characterized the scene of the murder to a greater or lesser extent. He might simply have a vague feeling that something evil occurred at that place. Or, he might be able to hold the murder weapon in his hand and mentally visualize the murder just as it happened.

Psychometry is, in fact, a way of traveling into the past. It is a form of time travel.

Although it appears reasonably certain that certain

individuals do possess psychometric abilities, the way in which psychometry works remains largely a matter of speculation.

According to one theory, living creatures and non-living objects exist simultaneously in the visible universe *and* in another, usually invisible universe in which time, in the usual sense, does not exist. Thus, if an event occurs, it does not occur in the past, present, or future; it simply occurs. In the visible universe, time erases the event. In the invisible universe, there is no time; hence, the event cannot be erased or even obscured by time. If this is the case, psychometry involves tuning in to the invisible universe and mentally reliving an event.

To many of the Indian tribes of North, Central, and South America, the concept of a timeless universe coexisting with the visible universe was not merely a theory—it was an integral part of their concept of reality. For instance, the Hopi language contains no verbs indicative of past, present, or future; only the duration of an event is measured.

To western man, time is visualized as progressing in a straight, horizontal line, from the past to the present to the future. The present is always becoming the past; the future is always changing into the present. Events that have already taken place are seen as forever frozen in the past.

To the Indian, however, there is no clear demarcation between the past, present, and future. Rather, time is seen as a continuum in which the past and future are inseparably intertwined with the present. The past is never gone forever. A dwelling once built is a part of the past existing in the present. Likewise, a tree once planted is a part of the past existing in the present.

It becomes apparent, therefore, that psychometry is not inconsistent with the Indian concept of time. On the other hand, it is somewhat more difficult to explain in the context of the western concept of time.

In any event, the possible implications of the widespread uses of psychometry are staggering. If it could be perfected to a high degree of accuracy, it would provide a means of traveling back through time without any danger of changing the course of future events. An event could be perceived as it actually occurred, but it could not be altered or prevented.

One of the important uses for psychometry would be in police work to help solve crimes in which sufficient clues cannot be found. Of course, psychometry may never be admissible evidence in a court of law, but it could nevertheless be used to locate stolen goods, kidnap victims, and fugitives. William and Elizabeth Denton, writing in *The Soul of Things* (1866), describe this possible use:

> It is a mistake, in more senses than one, to suppose that "dead men tell no tales." Their very bone cells contain the record of every deed committed and every outrage suffered. The dungeon hears and treasures every sigh of tyranny's victim, and the ashes of the martyrs contain the story of their wrongs.
>
> It has been said that the face of the murderer has been daguerreotyped [an early method of reproducing photographs] on the pupil of the eye of his victim Whether this is true or not, in the sense of ordinary vision, it is certainly [true] so far as psychometric vision is concerned.

Another important use for psychometry could be in solving some of the elusive mysteries of history. Could the riddle surrounding Hitler's disappearance or the assassination of John F. Kennedy be solved through psychometry?

These possibilities are explored by Hans Holzer in *Window to the Past: Exploring History Through ESP*.

In the book, Sybil Leek uses her psychometric talents to investigate the assassination of President Lincoln, and to explore the history of the *USF Constellation,* the arrest of Aaron Burr, and the first people to reach America by sea. Hans Holzer carefully recorded the words of Sybil Leek while she relived these events. Indeed, this is probably the most fascinating approach to the study of history that has ever been devised, and it may prove to be the ultimate cure for the student who is bored with the subject of history.

Following is another excerpt taken from *The Soul of Things* in which a piece of tufa (porous limestone) from the ancient Roman city of Pompeii* is being examined by a psychometer who *was not* told what the tufa was, nor where it originated.

> I see colored figures on a wall.
>
> Now, I see a yard, but it differs from all I ever saw before. It is diamond-shaped, with the corner in front of me. I feel the influence of something back of me; it seems like a building. It is a very heavy structure. I do not see it, but I feel its influence behind and on each side of me. This comes from some old country; the influence is ancient. It reminds me of Dickens' song of the Ivy Green,
>
> 'That creepeth o'er ruins old.'
>
> One side of this building looks out on the water; it may be the sea, for I feel the influence of some large body of water. The side of the building next to the water, and I think another side, have square towers rising up. I hear the rustling of long, heavy curtains in the building.
>
> In front of me, and to my left hand, the view is all shut out, and I have been trying for some time to find out the cause. It seems as if there was a great moun-

*The Roman city of Pompeii was catastrophically buried in lava following the eruption of Mt. Vesuvius, a volcano, in A.D. 79.

tain, so high that I have to elevate my head to see the top of it. The abrupt rise of the ground here seems to have caused that yard to be made in the peculiar shape it is. That mountain looks volcanic, and there are smoke, and stones, and cinders, and dust, all issuing from it in a dense body. They are thrown up with such force that, for a great distance, they form a perpendicular column, resembling somewhat a tall chimney, and then spread out on all sides. The mountain seems a hollow shell to a vast depth, the crater at the top being merely an orifice of small dimensions compared with the great cavern in the interior. The mountain has two peaks, the lower one much smaller than the other, but much sharper. I have been standing in the space between them, and now I go up a little higher. I hear the mountain bellow. What a depth that comes from! The influences that produce this eruption seem different from any I ever felt before. How strange it seems that I did not see this at first; for now everything seems so insignificant compared with it. The amount vomited out is immense. It is not like lava, but spreads out in a great black cloud, that rolls over and over and covers the country like a flood. I can hardly believe what I see is correct. It looks as if it would bury everything all around it. What a sight! There it goes, pouring, spreading, foaming, as it rolls down the mountainside in great black waves.

The narration continues with a description of the inhabitants:

I feel the influence of human terror that I cannot describe; it is awful. I see no one; but the feeling is almost overpowering. I feel like screaming. There are many different sensations commingled; but there is a horror more overpowering than all. This is either Herculaneum or Pompeii. There is no fancy about this; it is too terribly real. Some seem to regard it as a judgment of the gods. There is wild agony, prayer, and blind dread. Now I see them. Some wring their hands; others throw out their arms wildly. I see no

> one injured; that is, I perceive no impression to that effect.
>
> I feel the influence of some persons at a distance from here. Now, I see a very large crowd of persons, some hurrying along, and occasionally looking back; others seem to feel as if they could never leave, but are compelled to save their lives. The scene is agonizing in the extreme. I see one woman dart from the rest, and rush back, as if she had left a helpless child or parent to perish, that she was now determined to save; but she is compelled to give up in despair, for there is a fresh burst from the mountain, and she sees there is no hope. A darkness almost as great as night is now around them. How wild they seem! Many know not what to do, nor where to go. They act as if they thought there was hardly any place left in the wide world for them. There is a town at no great distance, to which many of them seem to be fleeing. I feel the influence dividing off in different directions; but I think many who escaped afterwards perished. Such is the impression I receive.

Though the subject matter in *The Soul of Things* was probably even more controversial in 1866 than it is today, it is apparent throughout the book that the Dentons were not fanciful individuals. On the contrary, in addition to being proficient in psychometry, they were keen and objective scientific observers. They carefully recorded observations gathered in the course of their experiments and, though they occasionally proposed possible theoretical explanations, they carefully avoided premature conclusions. This objectivity was evident in Elizabeth Denton's answer to a question, "But what is meant by 'going backward in time?' Do you wish to convey the idea that the spirit leaps, as it were, from the present to the past?" She responded:

> I do not assert anything in answer to this inquiry, save the sensation is in all respects that of closing the eyes on the present, and opening them on the past;

VIEW OF VESUVIUS FROM THE SITE OF POMPEII.

From *Pompeii and Herculaneum by* W. H. Davenport Adams.

MOUNT VESUVIUS PREVIOUS TO THE ERUPTION OF A.D. 63.

From *Pompeii and Herculaneum by* W. H. Davenport Adams.

MOUNT VESUVIUS AFTER THE ERUPTION.

From *Pompeii and Herculaneum by* W. H. Davenport Adams.

or, where the effort is made to do so, that of going from the present to the past by force of the will. But by what means this is accomplished—what is the cause of this sensation—whether the sensation be a true or false one, or what may be the philosophy of the whole matter, I can form no idea. I state only the fact, desiring that others may direct their attention to the subject, and arrive, if possible, at the true solution of the problem.

Elizabeth Denton began to receive psychometric impressions, even as a small child, before she understood their meaning or significance. To a large extent, she was naturally gifted with this ability. However, she was thoroughly convinced that receptivity to psychometric impressions could be developed by effort and practice. She offered several suggestions as to how this might be accomplished.

First, she explained that it is normal for many people to catch quick glimpses of patterns or objects while the eyes are closed and the individual is in a relaxed state of mind in a quiet, preferably darkened place. These images can be enhanced somewhat if a very slight pressure is exerted over the closed eyes with the fingers. At first, these images may come and go in a flash—so quickly, in fact, that one is not even sure what he saw.

By exerting a powerful effort of the will, and after much practice, these images can be made to stand still. After a time, they can be observed in considerable detail. When one reaches a point where he can clearly see these images, he will find that they are linked, in various ways, to his surroundings. For instance, one might reach into a box of old letters (in a completely darkened room) and hold one in his hand. An image might come forth of a face with distinct features. Upon further investigation, it might be ascertained that the face envisioned was that of the person who wrote the letter.

Thus, with considerable practice, one can develop the power of psychometry. The potential applications of psychometry are virtually endless. It can be used to answer questions relating to geology and archaeology. Using fossils and other relics of the past, one can obtain glimpses of ancient geologic events and early life forms. Applied to astronomy, one could use a meteorite and attempt to trace its origin and to follow it on its long journey through the heavens. Psychometry can be of immense value in diagnosing illnesses and in solving crimes. In fact, psychometry can be an invaluable tool to just about anyone.

10 Out-of-Body Experience

In the phenomenon known as astral projection (out-of-body experiences), visits to other dimensions are apparently quite common.

Astral projection involves the separation of the consciousness from the physical body in the absence of death. The experience may occur while a person is under general anesthesia, or after the ingestion of certain mind-altering drugs. It may also occur if a person is seriously ill or on the verge of death. At times, out-of-body experiences have occurred suddenly in normal, healthy individuals who had never had such experiences before. There are certain techniques by which one can learn astral projection, such as those mentioned by Oliver Fox and Robert A. Monroe. Also, it frequently occurs in the advanced stages of certain types of Eastern meditation. It should be noted that astral projection is not the purpose of meditation; nevertheless, it is a common side-occurence.

References to astral projection are found in many cultures, and some of these references are very ancient. It is mentioned in some very early Hindu, Egyptian, and Chinese writings. In general, wherever one finds a knowledge of magic and sorcery, one also finds a knowledge of astral projection.

In recent times, notable writers that have dealt with this subject include Oliver Fox, Sylvan Muldoon, Hereward Carrington, and Robert A. Monroe.

One of the most elementary forms of astral projection involves the sensation that one is looking down on his own body which is lying motionless beneath him. This type of experience has been described many times. For instance, Dr. Albert Hoffman, who first synthesized LSD, described looking down on his own body after he accidentally ingested some of the drug. Dr. Andrija Puharich had the same experience while merely resting one afternoon.

During astral projection, the consciousness may travel to virtually any physical location. One might visit a neighbor or a friend, or travel to the other side of the globe. Some individuals have described looking down on the earth from a great height. A certain Indian medicine man even claimed to have visited the moon several times in his astral form, and said that he was thoroughly familiar with its terrain.

In addition to traveling about in the known physical universe, the consciousness may also enter any of numerous, uncharted astral worlds that normally cannot be perceived at all. Eastern teachings allude to lower astral worlds, inhabited by evil entities, and to higher astral worlds, inhabited by advanced spiritual beings.

Robert A. Monroe, writing in *Journeys Out of the Body,* describes his encounters with an astral world which he says is a "physical-matter world *almost* identical to our own." He states that at first he thought it was some part of our world that he was unfamiliar with. Later, he concluded that "it can be neither the present nor the past of our physical-matter world." The society he encountered used no electricity and no electrical devices whatsoever, nor did they use internal combustion, gasoline, or oil. Nevertheless, they had locomotives and a peculiar type of very large automobile built to hold at least five or six people abreast. Apparently, time as it is known in this strange astral world does not in any way correspond to time as it is known in our own world. It is another realm with its own past, present, and future.

In addition to the "physical-matter [astral] world" described by Monroe, there are numerous astral worlds that exist only in a nonphysical, nonmaterial sense. They are inhabited only by incorporeal entities. There is, therefore, no sequence of physical events, no youth or old age, no passage of time.

Astral projection or out-of-body travel often leads to encounters with other realms that are far removed from the physical realities of everyday life. Time may exist in an entirely different sense, or it may be nonexistent altogether.

It is generally agreed upon that there are certain potential dangers involved in astral projection. Death can occur if for any reason one cannot return to his body. It is believed that spirits may intentionally attempt to sever the silver cord, in hopes of permanently gaining possession of the victim's body. It is also possible that one might encounter something so terrifying that it will have a lasting effect on the individual.

The dangers of OBEs (out-of-body experiences) are no secret, and have been readily acknowledged by authorities on the subject. However, it is also generally agreed that, if approached with caution, the potential benefits outweigh the potential hazards. It must also be remembered that some persons have little or no control over these experiences. They do not try to have an OBE; it simply happens.

I hope that I have given the reader sufficient background information concerning out-of-body experiences. It is futile, at this point, to elaborate on OBEs that involve unknown regions or dimensions. Far more relevant are OBEs which involve visits to remote places or other time periods *which can subsequently be substantiated.* As a case in point, I would like to cite the example of L. B. of California who started to have astral projections before she was old enough to understand them:

> I began having projections of consciousness and astral projections (involuntary ones) over twenty years ago, when I was a young girl. They frightened me so badly that I wouldn't tell anyone about them—I thought they would think I was crazy . . . In the first AP (astral projection), I found myself up near the ceiling of my bedroom, looking down on my inert body. In many of the early APs, I never left my room.

Describing a later experience, she writes:

> One night in 1948 I projected out of a deep sleep and floated into the front room. There I saw a burglar in a billed cap was trying to break in. This startled me back into the body. I now heard noises at the back door, and went into the front room and quietly phoned a neighbor . . . and told him someone was trying to break into my house.
>
> The neighbor came, but found no one. During the night my cat had kittens, and my neighbor swore that what had awakened me was the cat clambering up the door, trying to get in, before the kittens were born.
>
> I then showed him the splintered wood around the lock of the door—my proof positive that someone had tried to break in.

In other astral projections, L. B. was able to see things that had not yet occurred, which belonged not only to another place but to a future time period! Some of the things she saw dealt with terrible tragedies, notably kidnap victims and other persons who had met with foul play. In each case, she wrote down what she had seen—the condition of the victim and a detailed description of the surroundings in which the person (or body) was situated. She then sent this information to the Central Premonitions Registry in New York, an agency which files such information and, with the aid of computers, attempts to correlate this data with subsequent findings.

Newspaper articles that later appeared substantiated the things L. B. had seen, and proved to her (and to the Central Premonitions Registry) that her descriptions were amazingly accurate. In one incident, the victim was a young boy that had been brutally beaten and stuffed in a pipeline. In another case, the victim had been an eighteen-year-old girl that had been placed in a sleeping bag, bound with ropes, and left in a wooded area. In each case, the OBE occurred *at least a month* before the murder actually took place. In each case, L. B. had unknowingly travelled across the time barrier and witnessed a scene from the future!

L. B. has had numerous other OBEs involving other time periods, which are fascinating although they cannot actually be substantiated. In one such account, she writes:

> One night in 1967 I "projected" to the court of Louis XIV and saw a ball in progress—a beautiful sight. Everything was in miniature—it was like looking into a doll house. . . . The figures were not doll-like, but very human, with animated features and very graceful movements. All wore white wigs and rich costumes of the period in vivid colors.

Several books currently available describe methods by which one may learn the technique of out-of-body experiences. Paul Twitchell, a noted writer on the subject, actually offers a course which he teaches by correspondence based upon an ancient method known as Eckankar. Eckankar, he says, is a way to God-realization via soul travel. Twitchell differentiates between astral projection and soul travel, explaining that astral projection, though similar to soul travel, is a far more limited means of leaving the body. Study groups have also been formed in various places for students of Eckankar.

Fascinating research of a similar nature has been conducted by Professor Charles Hapgood and others using hypnosis. After the subject has attained a deep level of hypnosis, he is given a suggestion that it is a given time in the future, and asked to relate what he sees. For instance, the subject might be told, "It is now next Wednesday. You are holding the morning newspaper. What do the headlines say?" The method has met with considerable success. Many subjects are able to relate information that actually does come to pass just as they described it while in the hypnotic state.

Ironically, information obtained in this manner is far more easily verified than the better-known technique of hypnotic age regression, in which the subject is regressed through early childhood, into infancy, and finally into previous lives before his current incarnation. With certain notable exceptions (such as the case related in *The Return of Frances Willard* by Jeffrey Furst), hypnotic age regression findings are difficult, if not impossible, to thoroughly investigate.

Visits to the future through hypnotic suggestion are (at least theoretically) within the reach of any good hypnotic subject who has access to a skilled, open-minded hypnotist.

11 *Déjà vu*

Déjà vu is one of the most commonly described types of paranormal phenomena; many people have experienced *déjà vu* at some time during their lives. *Déjà vu* is the distinct sensation that what one is experiencing has happened before. This feeling often comes when it is least expected. Usually, the individual will rationalize that the event could not have happened before, and dismiss the whole thing as a funny feeling. He will not be able to remember it happening before, but the feeling is strong and unmistakable.

Déjà vu is certainly not a well-understood phenomenon. Some believe that it is merely a human quirk in which the normal thought processes become momentarily confused. Others, however, feel that *déjà vu* is a genuine psychic experience in which the mind temporarily enters into another dimension—a dimension in which time does not exist.

Another possibility is that certain events actually do repeat themselves in strange, peculiar ways. For instance, since 1840, every U.S. president elected in a year ending in zero has either been assassinated or has died in office. This pattern first began with William Henry Harrison, who was elected in 1840, and died in 1841. The only explanation for this phenomenon that I have ever heard is contained in an old legend. According to the legend, Harrison aroused the hatred of an old Indian medicine man, whose people had been slaughtered under orders of Harrison. Just before he died, the old Indian placed a curse on the U.S. presidency. Every president elected in a year ending in zero would fall victim to the curse.

Regardless of the reason, this pattern has repeated itself without exception up to the present day, as shown on the following chart.

President	Year Elected	Year of Death	Cause of Death
William Henry Harrison	1840	1841	died of pneumonia after one month in office
Abraham Lincoln	1860	1865	assassination
James Garfield	1880	1881	assassination
William McKinley	1900	1901	assassination
Warren Harding	1920	1923	stroke
Franklin Roosevelt	1940	1945	stroke
John F. Kennedy	1960	1963	assassination
?	1980	?	?

Even more uncanny are the similarities between the events surrounding the assassination of Abraham Lincoln and the assassination of John F. Kennedy.

Lincoln was elected in 1860; Kennedy was elected in 1960, exactly 100 years apart. Both died in office by an assassin's bullet. Lincoln had a secretary named Kennedy, and Kennedy had a secretary named Lincoln. Both were warned by psychics of their impending fates; both had premonitions of their deaths. Booth shot

Lincoln in a theatre and fled into a warehouse, while Oswald shot Kennedy in a warehouse and fled into a theatre. Both presidents were shot in the head, from behind; both had their wives present at the time; both were shot on a Friday. Booth was born in 1839; Oswald was born in 1939. Both assassins were killed before standing trial. Both presidents were succeeded by a successor named Johnson.

The list of similarities goes on and on. Lincoln and Kennedy were separated in time by an entire century, yet their lives seem to have followed parallel courses; the same events were repeated in a strange manner.

12 Drugs and the Sense of Time

It is a fact that mind-altering drugs usually alter the sense of time to some degree. In fact, this is one of the purposes for which psychotropic (mind-altering) drugs are sometimes used, both legally and illegally.

As a whole, cerebral stimulants tend to quicken the passage of time; an hour or a day seems to pass more rapidly. This is true of synthetic stimulants such as the amphetamines and Ritalin. It is also true of natural stimulants such as cocaine, from coca leaves. The Indian laborers in South America formerly chewed coca leaves to increase their endurance and to make the long day seem shorter. Some of the native Indian tribes of South America have used coca leaves for centuries, especially during tribal hunts when it is impractical to carry a bulky supply of food. The coca leaves serve as a valuable stimulant during long hunts, particularly in the rarefied air of the Andes Mountains. The Indians use a curious unit known as the *cocada,* which is the length of time and the distance a person could walk while under the effects of a mouthful of coca leaves. It is equivalent to approximately forty minutes, and varies in distance from two to three kilometers, depending on the terrain.

The highly stimulating leaves of *Catha edulis* (khat) are chewed in eastern Africa and neighboring regions to dispel drowsiness and to allay fatigue. It, too, makes time seem to pass more quickly.

The reader may remember occasions when a cup of coffee or two helped a slow morning to go by faster. Coffee contains caffeine, as do many other plants used throughout the world for their stimulating properties. These include tea (*Thea sinensis*), which originated in the Far East but is now popular in many parts of the world. Kola nuts (*Cola nitida*), which are native to West Africa, are used as a stimulant in Africa, and are also employed in the manufacture of cola beverages. Guarana (*Paullinia cupana*) and yerba mate (*Ilex paraguayensis*) are both native to South America and are used as stimulants in that part of the world.

In short, some type of stimulant is used in virtually every known culture to produce increased wakefulness and alertness, and to quicken the passage of time, especially during arduous tasks.

Perhaps the individual who is most acutely aware of the slow passage of time is the prisoner—the man or woman who is doing time.* Some years ago, when certain brands of over-the-counter inhalers contained Benzedrine, it became a common practice in certain prisons to crack open an inhaler and to dissolve some of the contents in a cup of hot black coffee. Those who did so claimed that it helped make time go by faster.

While stimulants generally accelerate the passage of time, depressants, as a whole, cause time to seem to pass more slowly (unless, of course, the depressant produces sleep). For instance, some individuals find it very difficult to work under the effects of antihistamines or certain tranquilizers, which often cause sedation. If sleep does not ensue, the day moves by at a snail's pace. This is also true of alcohol *after* the initial stimulation has worn off. It is sometimes true of marijuana, but marijuana is not purely a depressant.

*Time also passes very slowly for convalescing hospital patients.

Why do stimulants accelerate the passage of time while depressants have the opposite effect? Many psychological experiments have confirmed the observation that filled time passes more rapidly than empty time. In other words, time passes relatively quickly if an individual is busy and active. Stimulants tend to increase mental and physical activity. On the other hand, time passes relatively slowly if an individual is inactive. Depressants tend to decrease mental and physical activity.

The watched pot never boils, but if one puts a pot on the stove and then devotes himself to something else, the pot comes to a boil very quickly.

Hallucinogens alter the sense of time in a very complex manner; the effects produced vary according to many factors.

Those who have suffered through a bad trip have spoken of seconds that seemed like a long, grueling eternity. For instance, the noted philosopher Jean-Paul Sartre experienced only nightmarish visions after having taken mescaline. For him, the experience had been an unpleasant eternity.

On the other hand, those who have experienced the peaks of ecstasy while under the influence of an hallucinogen have often considered the experience to have been far too brief; they have spoken of the unpleasantness of returning so soon to everyday reality. A case in point is the experience of Peter Hurkos, who took the hallucinogenic mushroom *Amanita muscaria* in conjunction with a research project under the direction of Andrija Puharich.* He describes the experience in these words:

*_The Sacred Mushroom_ by Andrija Puharich, 1974. New York: Doubleday.

> I was in some far-off place of indescribable beauty. The colors, the forms are beyond description. The only way I can give you an idea of what I saw is that everything around me here is filthy, dirty, and horrible. It looks so ugly here that I hope you don't give me that mushroom too many times; I might not want to come back. . . . I know that where my mind was is real. Its beauty is beyond description. And this world is so ugly. I'm sorry I came back.

Puharich was very much surprised by Hurkos' experience, stating that it was very unlike his personality. "Peter is not the kind of person whose imagination would drift in the direction of rhapsodizing beauty," he said. "I have never heard Peter describe anything for its beauty's sake and enthuse over it."

In retrospect, then, an hallucinogenic experience often seems either too long or too brief. Even more revealing, however, is the way in which time is perceived while one is under the influence of an hallucinogen.

The person who has taken an hallucinogenic drug often feels that he has stepped beyond the normal barriers of time and space. He ceases to differentiate the past, present, and future, yet he becomes acutely aware of his own existence and the existence of things that surround him. He loses the ability to follow a logical thought process, and finds it impossible to recall what he was thinking or doing a short while earlier. He lives for the moment, yet he may be entirely oblivious to what came before, or what is to follow.

Time, as a concept, loses its significance. The sense of time is replaced by a sense of timelessness. The individual finds himself in a realm in which there is no time.

The late Aldous Huxley experimented with mescaline and, later, with LSD. Writing in *The Doors of Perception,* he states:

And along with indifference to space there went an even completer indifference to time.

"There seems to be plenty of it," was all I would answer when the investigator asked me to say what I felt about time.

Plenty of it, but exactly how much was entirely irrelevant. I could, of course, have looked at my watch; but my watch, I knew, was in another universe. My actual experience had been, was still, of an indefinite duration or alternately of a perpetual present made up of one continually changing apocalypse.

Tabernanthe iboga is an African shrub which contains the hallucinogen ibogaine. The cult of Bwiti in Gabon venerates this plant, and its members chew the root in a sacred ritual. James W. Fernandez, in an article entitled "Tabernanthe Iboga: Narcotic Ecstasis and the Work of the Ancestors," states:

> The timing of ritual activity is highly important, for the cult succeeds only insofar as it approximates the ritual of the other world. Alarm clocks are kept in most chapels to ensure that timings are exact . . . it is the impression of the initiates that they are involved in spiritual journeys for many hours, if not days.

An organization formed in Paris in the 1850s was known as the *Le Club des Hachichins,* and was devoted to the use of hashish, which is the highly concentrated resin produced by the marijuana plant. Hashish is many times more powerful than ordinary marijuana, and it is hallucinogenic in sufficient quantities. The club was composed of several noted literary figures of the past, including Gautier, Baudelaire, and Dumas.

Charles Baudelaire described his experiences with hashish in an essay entitled "The Seraphic Theatre." The following excerpt exemplifies the peculiar effect of hashish on the sense of time:

> But a new stream of ideas carries you away: it will hurl you along in its living vortex for a further minute; and this minute, too, will be an eternity, for the normal relation between time and the individual has been completely upset by the multitude and intensity of sensations and ideas. You seem to live several men's lives in the space of an hour.

Fitzhugh Ludlow, who experimented with hashish during the last century, describes his experiences in *The Hasheesh Eater.* Ludlow was able to obtain the drug legally from a local apothecary in the form of Tilden's Extract for about six cents a dose. Ludlow had taken the hashish extract on two occasions, but each time the dose was too small for him to notice any effects. A third time, he tried a larger amount. By then he had essentially concluded that he was unsusceptible to the effects of the drug, and he went to the home of a close friend.

He soon found himself in the company of several people, none of whom knew that Ludlow had taken the hashish. Suddenly, the effects started to become noticeable. A woman asked him a question in connection with a previous conversation. Concerning this experience, Ludlow wrote:

> As mechanically as an automaton I began to reply. As I heard once more the alien and unreal tones of my own voice, I became convinced that it was someone else who spoke, and in another world. I sat and listened; still the voice kept speaking. Now for the first time I experienced the vast change which hasheesh makes in all measurements of time. The first word of the reply occupied a period sufficient for the action of a drama; the last left me in complete ignorance of any point far enough back in the past to date the commencement of the sentence. Its enunciation might have occupied years. I was not in the same life which had held me when I heard it begun.

Later, he states:

> I left my room and went to the head of the staircase. The family had all retired for the night, and the gas was turned off from the burner in the hall below. I looked down the stairs: the depth was fathomless; it was a journey of years to reach the bottom! The dim light of the sky shone through the narrow panes at the sides of the front door, and seemed a demon-lamp in the middle darkness of the abyss. I could never get down! I sat me down despairingly upon the topmost step.

As the effects of the hashish were wearing off, Ludlow still retained an altered feeling about time:

> It was like returning home from an eternity spent in loneliness among the palaces of strangers. Well may I say an eternity, for during the whole day I could not rid myself of the feeling that I was separated from the preceding one by an immeasurable lapse of time. In fact, I never wholly got rid of it.

13 Can You See into the Future?

If you are one of those persons gifted with the ability to perceive an event before it actually occurs, you might wish to report your premonitions to one of the following agencies:

Central Premonitions Registry
Box 482
Times Square Station
New York, New York 10036

British Premonitions Bureau
TV Times
247 Tottenham Court Road
London
England W1P OAU

Similar agencies are appearing in all parts of the world. These agencies maintain extensive files of letters from people who have predicted future occurrences. Computers are programmed to weigh each prediction on the basis of the number of predictions received, the previous accuracy of those who have made the predictions, the frequency of incoming predictions, and other factors.

It is hoped that these methods will eventually be perfected to the point where catastrophes and natural disasters can be accurately predicted (and hopefully averted) before they actually take place. Even if the event could not be altogether averted, at least would-be victims could be evacuated from the sites of forthcoming disasters, and perhaps countless lives could be saved.

In reporting to a premonitions registry, give your name and address, and briefly describe the event you have foreseen. If the event takes place as you predicted, send in a newspaper clipping or other documentation, if available. If you demonstrate a high degree of accuracy, the registry may send you a questionnaire to fill out including such things as your age when you first had a psychic experience, whether or not you dream in color, whether you have ever had out-of-body experiences, and similar questions.

14 Premonitions–For What Purpose?

The most common premonitions that I experience involve telephone calls and letters. Though I may not be expecting a phone call or letter, I can usually discern in advance whether bad news or good news is involved. For instance, if the phone rings and I am about to hear a bit of bad news, I sometimes experience a generalized feeling of unpleasantness before I pick up the receiver. Likewise, if I receive a letter bearing bad news, I experience the same feeling of unpleasantness before opening the letter.

Sometimes I wonder whether premonitions simply happen, or if there are reasons for their occurrences. Should I respond to a feeling of unpleasantness by declining to answer the phone? Should I throw the letter bearing bad news away before opening it? A very close friend reminds me that "things don't just happen; they happen just. There is a reason for everything." I am inclined to agree with this viewpoint. I feel that in the ultimate structure of things, everything does have a reason. Sometimes the reason becomes apparent much later; sometimes not at all.

What, then, is the purpose of a premonition? Does it help to prepare one for something in the future so that it may be more readily accepted when the time comes? Many premonitions simply foretell an occurrence. The occurrence then takes place just as it was envisioned, and attempts to forestall or prevent it are utterly futile.

By contrast, some premonitions foretell tragedy—perhaps death or injury to a human life—and attempts to prevent the tragedy succeed. It would seem that such a premonition serves as a warning—a final opportunity to alter the future by acting in the present. Whether one attributes such things to God or merely to luck, they seem to be genuinely benevolent phenomena. It would seem that they serve no other purpose but to offer an opportunity to lessen human suffering. They do not actively intervene in anything, for one must heed the premonition in order to prevent an occurrence. Nevertheless, one is provided with an opportunity to avert a misfortune that could not otherwise be perceived.

One of the strangest premonitions on record, unique in its explicitness, is recounted by Flammarion in *Death and Mystery*. A woman awoke from a terrifying nightmare in which a ghost had held her arm while saying, "Either your husband or your daughter must die. Choose!" Haunted by the incident, she thought about the strange ultimatum from time to time. She concluded that if she were really given such a horrible choice, her maternal love might be dominant. Nevertheless, she dismissed the possibility that the choice would ever be other than hypothetical, because her husband and daughter were both in the best of health. A few days later, her husband died suddenly after a brief and undiagnosed illness. Fear was apparently not a factor, because she had not mentioned the dream to him.

This premonition seems, in retrospect, to have offered the woman a chance to avoid one of two gruesome tragedies. Fortunately, this premonition is not common or typical. But, like other premonitions, it offers the one who experiences it a last chance to avert something that may take place if no action is taken.

It remains an unresolved question whether these warnings originate from external sources—from God—

or from some unseen beings that are apparently concerned about the welfare of humans. Or are they a little-understood faculty of the human mind? Could it be that a part of the human mind is constantly in touch with the future? Is it possible that premonitions are the manifestation of an alarm mechanism that is triggered when a tragic event seems to be imminent? In other words, could it be that a subliminal knowledge of the future is always present in the human mind, becoming a conscious awareness only in the face of an impending emergency?

Regardless of the source of premonitions—external or internal—they involve at least a momentary dissolution of the barrier that normally separates the present from the future. That which is going to happen is perceived as though it were merely a stopping point on the continuum between the present and the future. The subject is somehow afforded a brief and limited peek into infinity. It becomes evident that past, present, and future are merely erroneous concepts—limitations of the human mind which cannot yet fully comprehend timelessness.

PART III:

Scientists and Writers Who Saw Through Time

15 Scientists Who Saw Through Time

If the ideas and accomplishments of an individual are not understood and appreciated until after his death, that individual is said to have been ahead of his time.

Indeed, there have been many people throughout history that fall into this category. Only in retrospect is it possible to truly fathom the contributions they have made.

Could any of these individuals have somehow obtained information directly from the future? Could someone from the past have traveled into the future and returned to his own time period with valuable information?

In the case of Nostradamus, described elsewhere in this book, there is little doubt that he somehow saw into the future and returned with a detailed account of the future course of world events.

On the other hand, has a scientist from the past ever traveled into the future and returned with information concerning an important discovery? Has an inventor from the past ever traveled into the future and returned to his own time period with the details of an important invention?

There is evidence that this has happened, although there is no conclusive proof. Nevertheless, the achievements of certain geniuses from the past have shown such enormous foresight that it seems that a definite glimpse of the future was involved.

Leonardo da Vinci (1452–1519) designed several machines that were apparently forerunners of modern military devices. One such device was an armored, four-wheel vehicle, capable of discharging "small stones in a manner almost of hail." Clearly, it was a precursor to the modern tank, though it was designed centuries earlier.

Da Vinci also designed various machines to save time and labor in the construction field. One of his inventions was a rudimentary bulldozer. It was powered by a large treadmill which in turn was powered by a number of men. The invention incorporated two large cranelike devices each attached to a huge bucket. It was calculated that the machine had a capacity of twenty wheelbarrows or 2,240 pounds.

Throughout his life, da Vinci was continuously fascinated with the idea of flight. It is said that he would buy birds from street vendors, only to immediately release them and carefully observe their movements as they flew away. He made numerous sketches of birds in flight, and studied them in great detail. He began to design flying machines. Some of his earlier drawings depicted machines that closely resembled modern helicopters. Later, he concentrated more heavily on what appeared to be gliders, and worked on ways to stabilize them in flight.

Finally, da Vinci actually built a flying machine, and his writings expressed his intense eagerness to test it. It was designed and constructed with the utmost precision and care. The day finally came when he did test it, but its maiden flight was apparently a failure. It is believed that da Vinci became very discouraged by this initial failure, and unfortunately abandoned the entire project.

It is ironic that most of da Vinci's inventions were not perfected and put into use until centuries after his death. His degree of foresight was uncanny; he was somehow able to visualize devices that were far ahead of their time.

Another scientist of the past who apparently possessed incredible foresight was Nikola Tesla (1856–1943). His accomplishments were equal to, if not greater than, those of Thomas Edison. He discovered alternating current (AC), and perfected the principles of modern radio. He invented fluorescent and neon lighting. He developed high-frequency currents and remote control.

In spite of his achievements, Tesla was never given the credit he deserves. Directly or indirectly, most of the modern applications of electricity resulted from the work of Nikola Tesla. Although we take his inventions for granted, many of us do not even know who Nikola Tesla was.

In *Prodigal Genius,* a biography of Nikola Tesla by John J. O'Neill, the author makes the following comment:

> Even as he walked among the teeming millions of New York he became a fabled individual who seemed to belong to the far-distant future or to have come to us from the mystical realm of the gods.

Many people first became aware of the possibility of a worldwide energy crisis during the gasoline shortage that began in late 1973. It was then that many people realized how dependent the world has become on fossil fuels such as coal, oil, and natural gas.

Incredible as it seems, Nikola Tesla actually confronted this possibility at least seventy years earlier. His conclusions and possible solutions to this question are perhaps best expressed in an article that appeared in *The Century Magazine* (June 1900, Pages 175–211). The article was entitled "The Problem of Increasing Human Energy, with Special Reference to the Harnessing of the Sun's Energy." The information it contains is as valuable today as it was in 1900, if not more so.

In reference to the conservation of energy, Tesla wrote the following:

> A similar inestimable advantage in the saving of energy available to man would be secured by obviating the great waste of coal which is inseparably connected with the present methods of manufacturing iron. . . . Though we are still far from the dreaded "exhaustion of the coal fields," philanthropy commands us to invent novel methods of manufacturing iron, which will not involve such barbarous waste of this valuable material from which we derive at present most of our energy. It is our duty to coming generations to leave this store of energy intact for them, or at least not to touch it until we shall have perfected processes for burning coal more efficiently. Those who are to come after us will need fuel more than we do. We should be able to manufacture the iron we require by using the sun's energy, without wasting any coal at all.

Tesla went on to develop an ingenious method of melting iron with electricity generated from moving water. He designed a hydroelectric generator to be situated beside a waterfall. The generator had the capacity to melt enormous quantities of iron, using no coal or other fuel. The immediate source of power is, of course, the moving water. However, the ultimate power source is the energy of the sun. The heat produced by the sun evaporates water from the surface of the earth, which later falls to the earth as rain and snow. Rain and melting snow then converge to form rivers, which, in turn, form waterfalls.

In 1900, most people considered the electric light to be one of the most advanced achievements of mankind. Tesla, however, was already considering ways to improve its efficiency:

> In our electric lighting systems we scarcely utilize one-third of one percent, and in lighting by gas a much smaller fraction, of the total energy of the coal. Considering the various uses of coal throughout the world, we certainly do not utilize more than two percent of its energy theoretically available. The man who should stop this senseless waste would be a great benefactor of humanity, though the solution he would offer could not be a permanent one, since it would ultimately lead to the exhaustion of the store of material.

Tesla was referring to the enormous loss of heat in the burning of coal to produce electricity. Today, much heat is still wasted in the burning of coal, oil, gasoline, and natural gas. For instance, a very large amount of heat energy is wasted by the modern automobile engine. After the heat is produced, it is merely dispelled into the atmosphere via the radiator and the exhaust pipe.

Tesla realized that, ultimately, man must obtain energy without consumption of any material. He considered such things as solar and geothermal engines, but came to the conclusion that these devices were quite limited in the amount of power obtainable, and in the rate at which they could generate power.

If Tesla had lived long enough, he would have undoubtedly discovered a means of harnessing the sun's energy that would have made the use of fossil fuels obsolete. In fact, there is evidence that Tesla may have made some highly advanced discoveries that were never disclosed. Some of his biographers (notably Margaret Storm) claim that Tesla invented an antigravity device.

He also claimed to have perfected a device that would render war obsolete. It would supposedly be able to generate an enormous force field around any nation, thereby rendering that country impervious to attack by land, sea, or air. The details of these inventions have never been made public.

There is no question that Nikola Tesla was many years ahead of his time. He predicted numerous scientific innovations that were not realized until years later. For instance, in 1926, he predicted the coming of television, when "we shall be able to witness the inauguration of a President or the playing of a World Series baseball game, just as though we were there." The public and the press laughed when he made this announcement. Today, television is very much a reality.

Writing in *Return of the Dove,* Margaret Storm stated that Tesla never designed his creations on paper, but that he had the power to mentally visualize them in perfect detail. She said that he could mentally envision a machine, leave it running, and then return to it later to examine the parts for wear! Clearly, Tesla's mind did not operate in this dimension only. Such powers of thought must necessarily belong to another realm or another dimension.

With respect to Tesla's search for new sources of energy, it should be remembered that we are still facing an energy crisis—more acutely than ever before. It is my opinion that we might have much to gain if the Federal Energy Office in Washington, D.C. took a long, hard look at the writings of Nikola Tesla.

Portrait of Nikola Tesla
from *The Century Magazine*, February 1894

16 Writers Who Saw Through Time

Many fictional books have been written that portray what life will supposedly be like in some future time period. One of the best-known books in this category is *1984* by George Orwell. *1984* focuses on the trend of increasing government involvement with the private lives of individual citizens and portrays a future society in which the government constantly monitors and controls the lives of its citizens. In essence, privacy has been done away with. Resistance to the government is combated with a combination of propaganda and brainwashing techniques.

There is no doubt that by 1984, the governments of most of the world powers will have the technology to monitor the lives of their citizens; perhaps the major governments of the world already possess that technology. Only a profound respect for the right of privacy can, in the final analysis, prevent the situation envisioned by George Orwell.

Another well-known fictional account of the future is contained in *Looking Back* by Edward Bellamy (1951). The book portrays a future society in which the labor inequities and the social injustices of the present have been brought to a successful conclusion. The change is brought about smoothly and gradually, with very little violence and social upheaval.

Looking Back is not actually a projection of current trends into the future; rather, it is a somewhat Utopian view of how much better things could be.

In addition to books which depict social and political situations of the future, there are literally hundreds of books in the category of speculative science fiction which portray future scientific innovations, and how they will affect people living in the future.

One noted author of speculative science fiction is Arthur C. Clarke, who is probably best known for *2001: A Space Odyssey*. In another of his works entitled *Profiles of the Future,* Clarke deviates from his usual fictional approach and limits himself to comments on possible scientific discoveries of the future. The book does not attempt to predict the future per se, but consists of a highly imaginative, though realistic, look at the world of tomorrow.

The types of books mentioned thus far either (a) portray the future as the author imagines it will be based on current trends, or (b) portray the future as the author wishes it would be.

There are also certain books that do not fall into either of these categories, yet they successfully predicted events that occurred long after they were published.

In 1560, a brilliant French physician, Michel Nostradamus, wrote a most unique work which amounted to a history of world events for the next 2000 years. Entitled *Centuries,* the book proved to be so highly accurate that Nostradamus has been called the only prophet in history who was never wrong.

It is not known what motivated Nostradamus to undertake this work. Early in the book, he gives somewhat vague reference to a method he used to project himself into the future; it consisted of a brass tripod, fire, water, and a rod. He may well have been deliberately vague in describing the device, however, because he was greatly shocked and dismayed over his visions of mankind's future. In fact, Nostradamus says that he rewrote *Centuries* after he had produced an initial version in which the names, dates, and places had been accurate in every detail. Fearing that the dreams and hopes of mankind might be permanently shattered if he revealed the horrors and sufferings of the future, Nostradamus wrote a revised version of the work in which the names and places were slightly altered, though not beyond recognition.

Probably the finest interpretation of Nostradamus' *Centuries* was given by the noted Dr. Manly P. Hall in a lecture he delivered on January 25, 1944.* The passages that follow are taken from that lecture, and are reprinted here by special permission.

> The predictions of Nostradamus concerning Napoleon are indicative of his general approach to the problem of prophecy. Remember that they were written about 1535 concerning events that would occur between 1750 and 1825. So we are dealing with an arc of over 200 years of time before these prophecies would come true. Let us see what Nostradamus said about Napoleon. (Remember that all of this material was written exactly as I am telling it. I have examined the original edition of the prophecies which was compiled in 1560.) Nostradamus says, "A man will be born *near* France. He will divide the armies of the French and will pass between them and become the leader of them; he will never be king but he will be proclaimed emperor; he will marry twice and have one son; this man will carry an eagle in his hand as the symbol of his authority, and his peculiar emblem will be two verticals and an oblique." (Two vertical lines and an oblique form the letter N.) "He will conquer nation after nation until the powers of Europe rise against him, defeat him, and exile him to a rock. Here he will remain for a time, then return to France

*"Nostradamus" by Manly P. Hall. A lecture delivered at Town Hall, Curran Theatre, Los Angeles, Tuesday, January 25, 1944. Arthur Linkletter, Chairman and Moderator. Copyright 1944 by Manly Palmer Hall.

and rule for a hundred days. After that, he will be defeated again and die upon a rock in the sea."

This is, in fact, an accurate account of the life of Napoleon, although it was written 200 years before the prophecies came true. Napoleon was born on the island of Corsica, which is off the coast of Italy and not far from France. He proclaimed himself emperor in 1804. He was married twice and had one son. After repeated military victories, his empire included Holland, Switzerland, and parts of Italy and Germany. Later, after being overwhelmed by the military strength of England and Russia, he abdicated as emperor in 1814, and was banished to the island of Elba. He escaped shortly afterward and raised an army, but was again defeated and exiled to the isle of St. Helena, where he died.

In reference to the prophecies of Nostradamus concerning Adolf Hitler, Dr. Hall says this:

> Adolf Hitler appears in the predictions. The nearest to the name of Hitler Nostradumus uses is Hister; he calls him "Hister, the Hysterical." He says that this hysterical one will be born not in Germany, but near Germany; that for nine years this one will keep the peace. In describing Hitler, Nostradamus uses the old Provençal word which has been translated as meaning "thin." Actually the word he uses was applied to a soup made entirely of vegetables and eaten during Lent. Thus he points directly to Hitler's vegetarianism. He describes Hister as a roarer and a bellower, one who spins stories that are not true, and the man who wrote "the book." He says that this man will take Vienna without firing a shot. He describes how this man will attempt to make himself master of the world, but that he will be removed from power or destroyed as a result of a series of circumstances taking place in the hills of Rome.

These predictions did, indeed prove to be accurate in every detail. Hitler was not born in Germany, but in Austria, which borders Germany on the south. "The book" he refers to was *Mein Kampf,* which became a sort of Bible to the Nazis, just as *The Communist Manifesto* is a Bible to the Marxists.

In reference to Nostradamus' prophecies concerning the period after World War II, Dr. Hall states:

> In order to accomplish the end of wars, we will have to solve two problems—one being that of the Far East and the other the Near East. . . . Nostradamus says that the great arc is moving gradually and inevitably toward the year 1997. He is very specific about that date which will be the time of a great war between the East and the West. That is when a united and consolidated Asia—including the Japanese islands, China, India, Siberia, and Mongolia—will form one great empire. Then it is that the great King of the East will rise and go to war in the air. And in this war flying machines will be so numerous and will move so continuously across the sky that the sun will be darkened as though by the flight of locusts, and death and destruction shall fall upon the earth. This war will determine the rulership of the whole earth. 1997 is the date.
>
> However, Nostradamus does not say that there is no possibility of avoiding this war, but he declares the avoidance of it demands and depends upon basic changes in human nature. Without those changes this great event cannot be prevented. He does not tell us the outcome of that war, except by indirection, but he goes on to say that after that war, the great prophet will rise in Christendom. The presbyter, or new priest, will reform and revitalize the entire structure of the Christian faith. We are led to assume from these statements that western civiliation will not be destroyed, because Nostradamus goes on to describe happenings in the West which imply that Asia does not conquer the Occident, although she would struggle greatly to accomplish this.

Indeed, Jeane Dixon, Edgar Cayce, and others have also predicted a great world holocaust that will make all other wars look minor by comparison. There seems to be close agreement that the event will occur at the close of the present century. Although a few prophets see this as the end of the world, most contemporary prophets regard this event as a major turning point in the history of mankind. Though the event itself will supposedly be accompanied by an enormous amount of suffering and bloodshed, it will be followed by peace and harmony and a new world order.

Perhaps we should not take Nostradamus' vision of the future too lightly. His batting average is nearly flawless. Also, it is highly significant that his predictions are in such close agreement with other prophets of the past and present.

Equally enigmatic was a man that appeared in the early part of the eighteenth century in the diplomatic circles of Europe. The man I refer to is the Comte de St.-Germain, although he masqueraded under many prominent names, and his exact identity has always been in question. He possessed seemingly limitless talents; he was, among other things, a scientist, historian, poet, artist, and musician.

The Comte de St.-Germain exhibited an intimately detailed knowledge of the history of the world for the preceding 2000 years. He explained that this knowledge was not due to study, but rather to memory—a memory of past centuries and the roles he had played in them. Likewise, he seemed to possess a prophetic knowledge of future events which astonished his contemporaries. For instance, he predicted the fate of the French ruling family several years before the outbreak of the French Revolution.

Though never understood, he was well-respected and famous throughout Europe. He was attributed with developing an elixir capable of prolonging beauty and youthfulness, whose efficacy was attested to by a lady of the French court. It is said that St.-Germain had the power to read minds, and could answer a question before it was ever spoken. It is believed that he could materialize and dematerialize at will, and could enter or leave a building without ever opening the door.

St.-Germain traveled extensively. He was encountered in Persia, Russia, Calcutta, Rome, Paris, and other places.

He was extremely knowledgeable in Eastern esoterics, and was an accomplished alchemist. Two of his works, *Studies in Alchemy* and *Intermediate Studies in Alchemy,* are still available.

It is not known where or when St.-Germain was born. He was supposed to have died between 1780 and 1785, yet he is reported to have been seen on different occasions as late as 1860. He spoke of the remote past as though he had been there and witnessed it in exacting detail. He spoke of the future as though it was already known to him. Perhaps he knew the secret of time travel; we will probably never know.

A Dweller on Two Planets by Phylos was first published in 1894. The book was actually written by a man named Frederick M. Oliver, who claims that an obscure occult master, Phylos, mentally dictated to him every word in the book.

Oliver describes his early life as having been carefree and lackadaisical, attributing this in part to the extreme leniency of his parents. At the age of seventeen, he became an assistant to Phylos, whose presence he could always feel when he wrote, although he could only see Phylos at certain times. Oliver says that he did not know what he was writing, and that he was merely an instrument of the true author, Phylos.

Phylos, speaking through Oliver, states that many of

the souls that inhabited the continent of Atlantis are now being reincarnated into the present-day world. As a result, he says, the world will make sweeping advances in science and technology. This will reflect the highly advanced technology of the ancient Atlanteans. The Atlanteans now being reincarnated are bringing much of their past knowledge with them.

It would be only too easy to discount Oliver as a deluded individual were it not for the fact that the book, which was copyrighted in 1894, *did* describe scientific inventions that did not appear until recent times. This could hardly be a mere coincidence.

In one section, he writes:

> We drew our electrical energies partly from the waves beating the ocean shores, more largely from the rise and fall of the tides: from mountain torrents and from chemicals; but chiefly from what might aptly be called the "Night Side of Nature."

He seems to be describing hydroelectric power when he refers to the generation of electricity from waves, mountain torrents, and so on. Electricity obtained from chemicals is the basic principle behind all storage batteries. However, the "Night Side of Nature" remains somewhat of a mystery unless, perhaps, he is referring to atomic power. Phylos said that "it was to our forces of the Night Side as an alcohol vapor motor is to thy steam engines." Apparently, the forces of the "Night Side" were far more powerful than any of the other forces known to the ancient Atlanteans.

In various parts of the book references were made to cathode rays. Phylos states that the people of Atlantis had the "power to traverse the aerial or marine depths" with "swift cars" and "subsurface sea ships." Later, he said:

> We could speak with, and see, and hear, and be seen by those with whom we wished to communicate, and this at any great distance and without wires, but over the magnetic currents of the globe.

Apparently, he was referring to radio and television communication. The "magnetic currents of the globe" probably refers to the fact that communication over long distances has largely been made possible by the presence of the Van Allen belts which encompass the earth, bending waves in a path that follows the curvature of the earth. But *A Dweller on Two Planets* was first published in 1894, whereas the Van Allen belts were not discovered until 1958.

In addition to technological achievements, the book mentions social laws that were unheard of in 1894. For instance, employees were paid according to a unique profit-sharing plan and periodic fluctuations in the cost of living were automatically taken into account. A government agency made quarterly estimates of the cost of living, and wages and salaries were adjusted accordingly. All employees were paid a fixed percentage of the company's net profits.

Phylos also said that Atlantean society provided absolute equality for both sexes; in fact, sex was considered immaterial with respect to voting, employment, and so on.

Lastly, in speaking of something that is still beyond the capacity of our own modern technology, Phylos remarked:

> Offer resistance to an electric current, thereby reducing or diverting it against an opposing force, and thou hast light; oppose to this (arc) light a combustible obstruction, and flame results. So mightest thou go on to the discovery soon to be made by the world of science, that light, all light, of the sun, or from any source, can be made to yield sound; upon this dis-

covery hinge some of the most astounding inventions thine age hath even dreamed of in its visions. But the primal discovery in this wonderful link, first of the sequence, will be the greatest of all, and so heralded. And this will be warranted, for the fact that it will be but a reincarnate unfoldment will not diminish its importance to mankind, nor the credit of its discoverer.

In 1898, a seaman and writer by the name of Morgan Robertson wrote a book about an enormous luxury liner more than 800 feet long, with a top speed of twenty-five knots, and a capacity of over 2000 passengers. The book, entitled *The Wreck of the Titan*, depicted a vessel built so well that it was considered unsinkable—virtually indestructible—by its makers. Thus, it was thought necessary to carry only twenty-four lifeboats. As the ship sped confidently through the fog during one voyage at about fifty feet per second, it came upon the unexpected—a gigantic iceberg. The ship was utterly destroyed and most of those aboard perished.

What is so uncanny about the book is the fact that the fictional *Titan* bore an uncanny resemblance to the *Titanic*, which was not constructed until 1911. Both were enormous liners. The *Titanic* was 882 feet long compared to the 800 feet of the *Titan*. Both ships had a top speed of about twenty-five knots; both ships had a capacity of over 2000 persons; both ships met their fates equipped with too few lifeboats in a sudden collision with an immense iceberg.

Though he may not have been aware of it, Robertson was describing an event that was to happen to a ship that had not yet been built. But this was not just another premonition of a future catastrophe that flashed across someone's mind; this premonition was *in print* when the catastrophe occurred!

The Secret Power was first published in England in

AERIAL-SUBMARINE VESSEL, ENTERING THE WATER.
Illustration from *A Dweller On Two Planets* by Phylos

AERIAL-SUBMARINE VESSEL DURING A NIGHT STORM.
Illustration from *A Dweller On Two Planets* by Phylos

1921, three years before the death of the author, Marie Corelli.

Marie Corelli is probably best noted for her ability to explore metaphysical subjects through the medium of the novel. She undoubtedly had psychic talents.

In *The Secret Power* she describes atomic energy and nuclear weapons with incredible accuracy. Since the book was published in 1921, it would seem that this reflects a genuine look into the future. Here is an actual quote from the original book:

> One man on the planet with power to destroy nations seemed quite a fantastic idea—yet science made it actually possible! He bethought himself of a book he had lately read concerning radio-activity in which he had been struck by the following passage—"Radio-activity is an explosion of great violence; the energy exerted is millions of times more powerful than the highest explosive substance yet made in our laboratories; one bomb loaded with such energy would be equal to millions of bombs of the same size and energy as used in the trenches. One's mind stands aghast at the thought of what could be possible if such power were used for destructive purposes; a single aeroplane could carry sufficient to annihilate a whole army, or lay the biggest city in ruins with the death of all its inhabitants."

In the story, Gwent (who possessed the knowledge of the "secret power") had intended to give it to one of the major world powers in exchange for the promise that it would be used on any country that attempted to wage war against any other country. Marie Corelli, speaking through the character Gwent, said:

> "I offer it to the United States—but I would have preferred to offer it to Great Britain, being as I am, an Englishman. But the dilatory British men of science have snubbed me once—and I do not intend to let

them have the chance of doing it again. Briefly—I offer the United States the power to end wars, and all thought or possibility of war for ever."

The incredible descriptions of the atomic bomb contained in *The Secret Power* seem as though they could have been written by an eyewitness to Hiroshima or Nagasaki—but they were written decades before.

Some years ago, a fascinating Chinese gentleman by the name of Gin Chow wrote a popular column for the *Los Angeles Times.* In 1932, at about the age of seventy-five, he wrote *Gin Chow's First Annual Almanac* (Los Angeles: Wetzel Publishing Co.). It is a delightful book containing an interesting autobiographical sketch of Gin Chow, weather predictions, and predictions concerning the future in general.

In describing his early life in China, Gin Chow tells about his grandfather who was a "great prophesier." In spite of the esteem in which he was held, his grandfather angered a Chinese lawyer who had his grandfather imprisoned until his death at eighty-four. Gin Chow's father also had a talent for prophecy, but avoided making predictions except to members of his own family lest he suffer the same fate as Gin Chow's grandfather.

Gin Chow goes on to describe his early struggle to make a living in California, primarily moving from one job to another. He also mentions his interesting experiences in love and marriage.

As for the future of the world in general, Gin Chow predicted in 1932 that hostility between China and Japan would not come to a complete end until 1947. It was in 1947, shortly after the end of World War II, that Japan actually did adopt a new constitution renouncing the right to wage war. This could be considered the year in which China was again safe from the harassment of Japan. Gin Chow spoke of a great world energy shortage that would eventually occur in which "there come no more wood in mountains an' no coal in ground. All fuel for fire burn up."

Apparently predicting the rise of the suburbs, he wrote:

> People will not live in crowded towns. Big towns not natural. Nobody likes um. Everybody live in little house of his own; maybe go to city to work but not live there.

This is suggestive of today's freeway commuters, isn't it?

Predicting the end of Prohibition a year before the law was repealed, he stated, "people get drink back but many get killed by driving automobiles while too drunk."

Clearly, many of Gin Chow's prophecies proved to be accurate in every detail.

PART IV:

Time Machines

Will man ever produce a time machine that could transport a human being to any past or future time period? If so, would such a machine also provide for a safe return to the present?

One of the chief arguments against this possibility is that no time travelers from the future have ever visited mankind. It is certain that such a visitor would have taken every precaution not to be noticed, but it is also probable that one of these visitors from the future would have inadvertently made some mistake, thus revealing his true origin. In the words of Arthur C. Clarke (*Profiles of the Future*):

> Though they might try to disguise themselves, accidents would be bound to happen—just as they would if we went back to imperial Rome with cameras and tape recorders concealed under our nylon togas. Time traveling could never be kept secret for very long. . . . chronic argonauts . . . would get into trouble and inadvertently disclose themselves.

But are we certain that this has never occurred? Are there no peculiar discrepancies in history that suggest the possibility that "chronic argonauts" may have visited the past? On the contrary, there is far more evidence that this may have occurred than most people imagine.

17 Fossil Footprints

Perhaps the most striking bits of evidence for time travelers are the fossil footprints and shoeprints that have been found which apparently predate the time when man first appeared on earth.

One of the earliest accounts of such a find was given by Henry Schoolcraft in the *American Journal of Science* (1822). He described human footprints that had been found near the Mississippi River at St. Louis, Missouri. Of course, there was no such thing as radiocarbon dating at that time, but the prints were found in crinoidal limestone, which was formed eons before man first walked the earth. Schoolcraft described the prints in this way:

> The prints are those of a man standing erect, with his heels drawn in, and his toes turned outward, which is the most natural position. The distance between the heels, by accurate measurement, is 6½ inches, and between the toes, 13½ inches, but it will be perceived, that these are not the impressions of feet accustomed to a close shoe, the toes being very much spread, and the feet flattened in a manner that happens to those who have been habituated to go a great length of time without shoes. Notwithstanding the circumstance, the prints are strikingly natural, exhibiting every muscular impression, and swell of the heel and toes, with a precision and faithfulness to nature, which I have not been able to copy. . . . The length of each foot, as indicated by the prints, is 10½ inches, and the width across the spread of the toes, 4 inches, which diminishes to 2½ inches at the swell of the heels, indicating, as it is thought, a stature of the human size.

After carefully examining and scrutinizing the prints, Schoolcraft concluded:

> Every appearance will warrant the conclusion that these impressions were made at a time when the rock was soft enough to receive them by pressure, and that the marks of the feet are natural and genuine.

In 1882, O. C. Marsh reported the discovery of human-like tracks excavated near Carson City, Nevada. The tracks are smooth, apparently resembling sandal prints.

Then, in 1896, the following item appeared in *The American Anthropologist* (vol. 9, page 66):

> It is reported that H. E. Huford, of Kemper Lane, Walnut Hills, Cincinnati, exhibited before a recent meeting of the Ohio State Academy of Science, a large stone taken from the hillsides four miles north of Parkersburg, on the West Virginia side of the Ohio River, about twenty years ago, *in which there was the imprint of a perfect human foot,* 14½ inches in length. [Italics are mine.]

Several similar finds were made more recently in Colorado, Arizona, Utah, and other parts of the United States. One of the most interesting specimens was reported in 1968 by Dr. Clifford Burdick. It was discovered at Antelope Springs, Utah, and has the appearance of a modern shoeprint with the heel and sole clearly defined. It is about 15 million years old, dating back to the Cambrian Period.

All of these finds are utterly baffling to archaeologists. They do not seem to fit into a logical pattern; they seem to be completely misplaced in time.

It has been suggested by some that these human-like footprints may have been made by extraterrestrial visitors that came to earth eons ago. There are, however, inconsistencies in this theory. If the prints were made by extraterrestrial visitors, we must assume that these aliens just happened to have feet that were the same size and shape as those of a normal human being. And, at least in one instance, the alien must have been remarkably accustomed to the earth's atmosphere to walk around wearing ordinary shoes rather than any type of protective spacesuit.

Perhaps a more plausible theory is that some of these anomalies were made by visitors from a future time period. If a time traveler deliberately chose to visit earth of 15 million years ago, there would be very little danger that he might inadvertently do something that would radically alter the course of history. This should, of course, be a prime concern to any future time traveler. Even with the best of intentions, he might perform an act that could have serious, unfavorable consequences. In the words of Arthur C. Clarke (*Profiles of the Future*):

> The classic argument against time travel is that it would allow a man to go back into the past and kill one of his direct ancestors, thus making himself—and probably a considerable fraction of the human race—nonexistent.

Or, a time traveler to the past might prevent one of the tragedies of history, only to make way for another, unforeseen tragedy. Probably the safest time period to visit would be the period predating man's appearance on earth. At least, it would minimize the possibility of altering the future course of history.

18 Enigmas of History

Equally puzzling are the unexplained cases of people who have seemingly appeared out of nowhere, speaking strange tongues and describing unknown places.

For instance, medieval chronicles describe a peculiar incident that happened in England during the twelfth century. Two children, a boy and a girl, were found by reapers in a field near some wolf pits. The two children were green in color, wore strange clothes, and could not speak. At first, neither the boy nor the girl would eat anything that was put before them. According to one account, they were finally brought some raw beans which were torn from their stalks, which they both grabbed with considerable eagerness. Oddly, they seemed to look for the beans in the stems rather than in the pods, and were very disappointed when they could not find them. Someone nearby opened the pods for them and showed them the beans, which they ate ravenously.

The boy died some time later, but the girl gradually became accustomed to her new environment, learned the language, and lost the green coloration in her skin. Upon being asked where she had come from, she replied that she came from a land that was continually suspended in twilight. It had, she said, no direct sunlight, but it was not completely dark either. She said that her people could see a land in which there was light, but that they were separated from it by a very wide river.

Although the girl managed to adjust to her new culture, the mystery of her strange origin was never solved.

In a later incident that occurred in May 1828 and is recounted in *Caspar Hauser: The Enigma of A Century,* by Jacob Wasserman, a strange youth judged to have been about sixteen or seventeen wandered into the German town of Nuremberg. He could only speak two sentences in German, and was unable to state who he was or where he had come from. Upon being taken to a police station and given a pencil and paper, he wrote the name Caspar Hauser.

Caspar Hauser mystified the residents of Nuremberg because he seemed completely unfamiliar with common objects and ordinary, everyday activities. He looked at everything as though seeing it for the first time. It is recorded that he once watched a candle with apparent fascination, and then attempted to pick up the flame. He was, of course, badly burnt.

In 1850, a mysterious stranger wandered into a small German village, who said his name was Joseph Vorin. He said he had come from a place called Laxaria in Sakria, which was far from Europe and beyond vast oceans. Actually, no such places as Laxaria or Sakria are known to have ever existed. Could Joseph Vorin have been a visitor from another plant? Another dimension? Or a time period in the distant future?

Suppose a man named John Smith from New York City were suddenly transported back to China of 100 B.C. What would the ancient Chinese think of the strange, fair-skinned man who spoke an alien tongue? And if, by some remote chance, they could decipher his strange language (modern English with a New York accent), what would they think of his description of New York City? They would understand him no better than our ancestors understood Caspar Hauser or Joseph Vorin.

If it will someday be possible to travel into the past, it might also be possible to travel into the future. Clearly, this would have both advantages and disadvantages. It would enable the time traveler to see all the good that the future holds, but all the evil as well. At least, it would eliminate uncertainty, but this is sometimes—not always—desirable. For instance, a person who has a loved one in the armed forces classified as Missing In Action is painfully tortured by uncertainty. He yearns to know whether the loved one is dead or still alive. He could accept the worst, but he cannot accept the pain of not knowing.

On the other hand, would anyone want to see the tragedy and suffering that the future inevitably holds? In other words, can man live without his dreams and hopes? Perhaps one man who saw into the future has already faced this dilemma; namely, Nostradamus. In a lecture delivered at Town Hall in Los Angeles, January 1944, Manly P. Hall stated:

> Nostradamus says that after he had completed his prophetic *Centuries,* he saw a pattern or picture of world events so complicated and so awe-inspiring, so enormous in terms of human suffering and human misfortune, so horrible in the terms of things that human beings would do to each other in the name of civilization and progress, that he wondered whether it was wise to reveal to the world the misfortunes that lay ahead. Would man's initiative, his hopes, his dreams, be shattered if he realized the terrible things that he would have to go through in order to find the peace and security that he was seeking? Nostradamus came to the conclusion that it was not good to reveal too clearly these things that might too profoundly influence the course of human living. So he rewrote the prophecies, obscuring them, veiling them, so that only

scholarship could reveal the meaning. He said, 'In many instances it will be impossible to interpret my prophecy until the event itself occurs. Then you will discover that I have placed within the prophecy somewhere a key, a peculiar, limiting circumstance by which it is possible to perceive that this event, and this event only, is the fulfillment of it."

19 George de la Warr

Until his death in March 1969, the late George de la Warr conducted extensive research at his laboratories in Oxford, England, on the properties of subtle types of radiation emitted by both animate and inanimate objects.

According to de la Warr, writing in *New Worlds Beyond the Atom,* all things constantly emit electromagnetic radiation of characteristic frequencies. This is true for simple elements and molecules, as well as for both lower and higher forms of life. With respect to living creatures, a healthy organ such as the heart emits a characteristic radiation pattern. If the organ is diseased, it emits a different radiation pattern, which varies according to the nature of the disease. Likewise, a three-month human embryo emits a radiation pattern that is somewhat different than a six-month human embryo.

One of the results of de la Warr's research was the development of a "thought resonator" (or "cavity resonator") which enables the operator to tune the radiation emitted by his own mind so that it corresponds exactly to that emitted by the object that is being investigated.

The device consists of a box with a resonant cavity and a metal plate which also acts as a resonator. An "antenna wire" connects the box with the object under investigation. There is a thin rubber diaphragm on top of the instrument. The operator concentrates on the object he is investigating, and rubs his finger over the diaphragm. When there is resonance*, a distinct crackle and sticky feeling is experienced which is referred to as "stick."

Langston Day, writing in *New Worlds Beyond the Atom,* describes the phenomenon as follows:

> If we follow up what is happening and why this stick proves that there is resonance, we shall have to cross the frontier which divides the world of matter from another world where mind is an operative force. For as we shall see later, the mind acts as a modulator of complex and continually changing wave-forms** which are emitted from the human body. To hold in mind a steady thought is to attune your bodily radiations to those of that something.
>
> [In the case of a plant,] to tune-in to the plant's radiations the operator focuses his mind on the plant. The wave-forms which he is emitting are then in a harmonic relationship with the plant's radiations and this reinforces the effect. If there is resonance . . . it is revealed by a crackle in the rubber diaphragm when the finger is passed over it.

De la Warr subsequently made the astonishing discovery that a leaf emits the same radiations that are emitted by the plant from which it was severed. Similarly, a blood sample or even a photograph somehow retains a link with the person from which it is obtained, and emits the same radiations.

De la Warr found that it was possible to diagnose diseases through the use of a blood sample or photo-

*resonance: the existence of a harmonic relationship between two or more different wave-forms. In music, two or more harmonious notes combine in resonance to form a chord.

**wave-form: any of the numerous forms of electromagnetic radiation including light waves, sound waves, cosmic rays, and the more subtle radiations that were investigated by de la Warr. Each wave-form has a specific wave-length.

graph connected to the thought resonator, although the patient might be hundreds of miles away.

De la Warr obtained even more astonishing results from a special camera he developed based on the principle of the thought resonator. Under the right conditions, he found that he could actually photograph a diseased organ from the radiations emitted by a blood sample taken from a patient or his photographic likeness. The resulting picture looked very much like an X-ray picture clearly revealing the nature of the abnormality. In separate experiments, he was able to photograph tuberculosis of the lungs, ulceration of the stomach, and a malformed fetus. When the camera was connected to a blood specimen obtained from a cow, the resulting picture revealed a piece of wire and a stone lodged in the cow's stomach. Subsequent surgical removal confirmed the diagnosis.

Perhaps by now, the reader justifiably wonders what all this has to do with time travel. In other experiments, de la Warr found that the radiations emitted by objects do not always concern their present state, but may also reveal phenomena of the past or future. For instance, a photograph taken of a lily seed revealed an image of the bulb of a mature lily plant; another photograph revealed the flower of a lily plant. The photographs were somehow displaced in time. They did not show the lily seed in its present form. Rather, they revealed that which the seed could potentially become. Similarly, a photograph taken of some homeopathic *Aconitum napellus* pilules revealed an image of a mature *Aconitum napellus* plant.

Realizing that his camera was apparently not bound by time, de la Warr attempted another experiment in which he took a vial of his own blood and a vial containing his wife's blood. After making certain adjustments, he concentrated intensely on one thought: My

Wedding Day, 1929. After developing the photograph, the resulting picture revealed two forms resembling human beings, presumably de la Warr and his wife on their wedding day.

In many ways, the experiments of George de la Warr were crude and rudimentary, leaving many questions unanswered. Nevertheless, he did open up a whole new realm of scientific inquiry. The subtle radiations he investigated are not really a part of the physical world. Again quoting Langston Day (*New Worlds Beyond the Atom*):

> Resonance is not in the three dimensions of space, nor is it in time. It appears to belong to another dimension which introduces a principle of unity or connectedness that is not present in the four-dimensional continuum.
>
> Things resonate because they are in a certain harmonic relationship, not because they are simultaneous or contiguous. Hundreds of miles may lie between them, nevertheless there is this strange connection.

Today, the work of George de la Warr is being carried on by the Delawarr Laboratories Ltd. in Oxford, England.

20 The Hieronymus Machine

The findings of George de la Warr have been curiously paralleled by the work of an American radio operator, electrical engineer and inventor by the name of T. Galen Hieronymus. In the early part of World War I, he worked on the development of a wireless telephone. During 1919–20, he researched new methods of radio broadcasting as a member of the Army Signal Corps with the Rainbow Division in France.

Hieronymus' interest in radionics began when a neighbor, Dr. Planck, asked him to machine various parts for a very peculiar electronic device. Dr. Planck attributed the device to a brilliant medical genius involved in new methods of treating disease, but did not go into further detail. Subsequent experiments showed that the machine was somehow capable of picking up emanations from both animate and inanimate objects.

Hieronymous continued to experiment with and modify the machine, and to try to determine its *modus operandi*. Essentially, the machine worked on the same principle as de la Warr's thought resonator. The operator tunes his mind to the object in question, and rubs his fingers across a rubber diaphragm until it "sticks." In this way, one can pick up emanations from diseased and healthy tissue, minerals, and other living or nonliving things.

What has this to do with time travel? The Hieronymus Machine can be adapted to photographically record things of other time periods. For instance, prehistoric creatures have been photographed from the radiations emitted by ancient fossils! In this way, de la Warr's thought resonator and the Hieronymous Machine are also time machines, actually capable of recording scenes from the past and future. In effect, they capture on film what the psychometrist captures in his mind.

Hieronymus applied for a patent on his machine in 1946, and in 1948 he was granted U.S. patent 2,482,-773. Like all U.S. patents, a copy may be obtained from the U.S. Department of Commerce, Patent Office, Washington, D.C. 20231, for a nominal fee.

In an interesting experiment which demonstrated the far-reaching capabilities of his machine, Hieronymous obtained photographs of each of the three Apollo 11 astronauts and inserted them, one at a time, into his machine. In spite of the enormous distance involved, he was able to monitor their physiological functions during the entire mission. He later found that the data he had gathered corresponded closely to the official data obtained by NASA. However, Hieronymus obtained an important bit of additional information that escaped the NASA scientists. He stated:

> Of all the data collected . . . the most important and startling is that *there is a lethal radiation belt on the Moon,* apparently extending from about sixty-five miles down to approximately fifteen feet from the surface. . . . There was a noticeable drop in the general vitality [of the three astronauts] and an increase in the carcinogenic readings. The pathologies increased until the men had actually stepped on the Moon, then everything reversed.

After monitoring subsequent moon landings, Hieronymus concluded that the lethal radiation belt, though always present, may vary somewhat in its exact location but that further studies would be necessary to confirm this hypothesis.

In other investigations, it was found that the Hieronymus Machine could be adapted to effectively eradicate insect pests from infested plants, without harming the plant in any way. It was also found that this could be done at a considerable distance without any loss in efficiency.

This experiment, though designed for the betterment of mankind, has raised serious questions concerning the inherent dangers of the machine if used for evil purposes. It has been pointed out that the machine, like atomic energy, has immense potential. It can truly improve the lot of mankind but, used improperly, it could cause great suffering and harm.

Hieronymus is aware of the enormous capabilities of his discovery, and has been understandably hesitant to reveal all that he knows. Ultimately, however, mankind must resolve to use the resources at his disposal for benevolent and peaceful purposes, or face the possibility of self-extinction.

21 The Project Hermes Device

The Borderland Sciences Research Foundation (BSRF), which is headquartered in Vista, California, was founded in 1946 by the late Meade Layne, and is currently under the direction of Riley Hansard Crabb. The purpose of the organization is perhaps best expressed in the words of Meade Layne in reference to the *Journal of Borderland Research,* the society's publication:

> BSRF publications are scientific in approach, but employ few technical expressions. They deal with significant phenomena which orthodox science cannot or will not investigate. For example: the Fortean* falls of objects from the sky, teleportation, radiesthesia, PK effects, underground races, mysterious disappearances, occult and psychic phenomena, photography of the invisible, nature of the ethers, and the problem of aeroforms (flying saucers).

The BSRF does not endorse any particular device used in psychic research, but rather serves as a clearinghouse for novel and potentially productive ideas in this area. One of the most unusual and promising devices publicized by the BSRF is the Project Hermes Device, an electronic machine used to stimulate the ESP center in the brain through the use of a weak, controllable, oscillating field. The Project Hermes

**Fortean* is derived from the name of Charles Fort (1874–1932) who investigated paranormal phenomena.

Device should not be confused with the alpha-wave detectors that have recently appeared on the market, which simply detect alpha-type brain waves.

Design specifications for the original Project Hermes Device were supposed to have been received telepathically by a man named Ralph Holland prior to its release in 1951. Holland described the device as follows:

> The device is supposed to be a reproduction of an ancient communication instrument. . . . Its main use by the ancients was to communicate telepathically with others on [this plane of existence]. My informant says, however, that it can be used to communicate with entities on other planes, if properly tuned. This particular form is very short range under ordinary circumstances; that is, its effect will influence only persons or entities in the immediate vicinity, when used singularly. If, however, two persons have such a device, and the devices are tuned to each other, they will work over fairly long distances. The actual method of operation is to create a controlled vibrational field which will sensitize those within its influence.

Some time after the details of the Project Hermes Device were first presented to the BSRF, reports started coming in from independent researchers who had built the device and experimented with it. Nearly all of the researchers had reported achieving no results whatsoever. By contrast, one investigator, Ed John, reported a high degree of success:

> This device is a sneaker. It works and works well. Perhaps too well. I have discovered that it isn't wise to use it too often, as it tends to stimulate some unused portion of the brain, and causes kickbacks of some rather strange psychic phenomena, even semi-materializations of objects that happen to be uppermost in the operator's mind at that particular time of use.
>
> I have already tested it on many others and it has not yet failed to make some sort of response on every

> test person. Some experience great clairvoyance. Others see beings in other worlds, *and still others see and hear both backwards and forwards in time.* [Italics are mine.] Others merely see the souls of those who have died some time ago.

Ed John alone was able to obtain successful results with the device. After this rather sluggish beginning, BSRF received little noteworthy correspondence concerning Project Hermes for some time. Then, in 1964 and 1965, two investigators, Joseph Ponciano and Ray Hudd, worked together with the device, and later provided BSRF with eighteen pages of notes and data on their findings. Basically, their conclusions were in agreement with those of Ed John. In the words of Joseph Ponciano:

> This research, testing, and experimentation will run into years before any definite conclusions can be drawn about this electronic ESP device. Most persons didn't receive anything on their first attempt; it was only after the second or third session that most participants started to receive impressions. The ESP results in these sessions seemed to be cumulative in that the more sessions a person participated in, the easier it was to stimulate him. In the brief number reported here, the types of ESP experienced are: clairaudience, clairvoyance, telepathy, precognition, and retrocognition.
>
> The aftereffects noted have been minor. Some experience beautiful flower smells or other exotic odors during the testing. A few have noticed these strongly the day after the sessions.

The following is an account of one subject's experience with the Project Hermes Device. It appeared in the *Journal of Borderland Research* (vol. 21, #5, July-August 1965), and was submitted to that publication by Joseph Ponciano and Ray Hudd. It is reprinted here with the kind permission of Riley Crabb, director of the Borderland Sciences Research Foundation.

> After I put the device on, my vision was taken to an old Victorian house of several stories high, which was perhaps in the 1800s. It was properly cared for, the grass and so on. I recall there was a red carpet down the steps and saw horses and carriages. Time rapidly passed and I saw the red carpet go away. I saw the house become grey and go down to an older stage. Then I was taken on the inside up the stairs and to the attic. On complaining about it being very dark, they opened the door and the attic was filled with light. In the attic was information of some kind, which at the moment I don't know. I felt it would come at a later date.

There are many obscurities in the above account, but it does illustrate the apparent ease with which the mind of the subject can perceive things on another level of consciousness. The Victorian house was first seen in its prime, and was then seen in the process of aging and decay. The subject's mind was apparently able to scan a process that took place over many years.

In summary, the findings of Ed John, Joseph Ponciano and Ray Hudd seem to indicate that a relaxed, receptive attitude is most conductive to a fruitful session. If the subject is skeptical from the start, he may inadvertently block out impressions. Likewise, if the subject tries too hard to receive impressions, he may defeat the results. It would also seem that the more sessions a person participates in, the more likely he will be to be stimulated by the device.

A word of caution. Even if impressions are received with the device, it is advisable to use intelligent discrimination with respect to the material received. Many mediums believe that information obtained from other dimensions or planes of existence is as likely to be in error as information obtained in the everyday world. This point was well-expressed by Hudson Tuttle in *Mediumship and its Laws* (1900):

> A spirit, simply because a spirit, cannot prophesy. The greater proportion know little of cause and effect and are incapable of hazarding more than a conjecture. Yet these ignorant spirits are first to prophesy future events, and bring shame to those who trust them.

As stated before, the BSRF does not make any claims for the Project Hermes Device, nor do they endorse it. They do feel, however, as I do, that it is a most interesting instrument worthy of further research.

The BSRF offers a brochure entitled "Two Inventors Return" which contains instructions for building the Project Hermes Device. At the time of this writing, the brochure sells for $2.00. (California residents add current sales tax.) It is available from: Borderland Sciences Research Foundation (BSRF), P.O. Box 548, Vista, California 92083.

22 The Integratron

Situated about seventeen miles outside of Yucca Valley, California, is the Integratron. The massive, domelike structure, which is about eighty-five percent completed, is a project of the College of Universal Wisdom, which is under the direction of George W. Van Tassel. When completed, the Integratron is supposed to act as a time machine, making possible rejuvenation of the aged, prolongation of youth in the young, antigravity, and time travel. According to Van Tassel, the only thing standing in the way of its completion is the need for additional financial support from the public, which is the project's sole source of income.

George Van Tassel was born in Jefferson, Ohio, in 1910. He moved to California in 1930. He worked in the aircraft industry until 1947, when he leased Giant Rock, an abandoned airport, from the government. In 1951, Van Tassel purchased the adjoining forty-acre ranch.

Van Tassel recounts an incident of many years ago when he was taken aboard an extraterrestrial space ship that had landed in the desert. He was given a tour of the vessel, and was presented with information of a very advanced nature that made the construction of the Integratron possible. He wrote about his experiences in *I Rode a Flying Saucer,* which appeared in 1952. Since that time, Van Tassel has had other contacts with extraterrestrial beings. He has devoted much of his life to the Integratron and related research.

Giant Rock Airport takes its name from an enormous, sixty-foot granite boulder under which Van Tassel has built a most unusual study and personal library. Until her recent death, Mrs. Eva Van Tassel managed a cafe situated near the study and not far from the Integratron.

Van Tassel presides over an annual Space Convention held each year at Giant Rock Airport, which always draws enormous crowds. At these conventions, people from all over the world come to share their experiences in which they have contacted or seen space ships or beings from other worlds.

The Integratron itself has the appearance of a large, white dome encircled by small rectangular windows. The plan for the Integratron requires that no metal be used in the outer structure; consequently, it has been assembled without any nails or screws.

Inside the dome, a self-energizing machine is under construction which will create a powerful magnetic field to act on the cells of the body. Van Tassel likens the cells of the body to miniature batteries, and he says the Integratron will be able to recharge them "as you would any battery."

How does the Integratron relate to time travel? First, it is necessary to treat briefly Van Tassel's concept of time, which is admittedly a difficult one. His concept of time enters into his concept of God. He believes that God is composed of a trinity of time, space, and being. Time does not pass by. Rather, all that moves in the universe passes through time. Time consists of infinite lines of force. When a person or other object moves, he opposes these force lines (passes through time). Man erroneously measures the passage of the earth through time and supposes that "time goes by."

Time seems to pass at variable rates to each individual because of the rate of his *own* passage through time. He may move through time at a faster or slower rate; thus, time *seems* to pass quickly or slowly. The effects of time are, therefore, most noticeable following heavy physical exertion. They are least noticeable following sleep.

If a person ceased all movement so that even the cells in his body were utterly motionless, he would be moving through time very slowly, but he would still be moving through time because the earth itself would still be in motion. The motion of the earth consists in its revolvement around its axis and its orbit around the sun.

The magnetic forces of the earth generate a time field around the earth. If another, artificial time field could somehow be generated, it could lessen or modify the effects of time on a human being or other living creature.

The Integratron not only generates a powerful magnetic field to recharge the cells of the body; it simultaneously generates its own time field. Riley Crabb (who is not affiliated with the College of Universal Wisdom) describes it in this way:

> We'll put it crudely like this. Electricity flowing along a wire is one dimension. It creates a magnetic field around the wire, two dimensions. The magnetic field creates an electrostatic field, three dimensions. Finally, the electrostatic field creates a tempic [time] field and this is the fourth dimension.

In other words, the magnetic forces in the earth generate a time field around the earth. The Integratron, by generating a time field in opposition to the earth's time field, will be able to reverse some of the effects of time.

Thus, the Integratron should theoretically be able to double or triple a person's lifespan. Van Tassel says that it will rejuvenate those who are showing signs of aging, as well as prolong youth in those who are still young. When completed, the machine will be able to handle as many as 10,000 people a day.

Van Tassel says the machine has already been tested on mice, and they came out healthy and invigorated. Their reproductive capacity, he says, was enhanced, and they lived substantially longer than normal.

23 Crystal Gazing

Today, crystal gazing is usually thought of in the context of carnivals and fairs. To some, it is a way to have fun. To others, it is simply a waste of money. At any rate, it is seldom taken very seriously.

Crystal gazing, as it is usually practiced, should not be taken seriously at all. One should not be concerned over what he hears at a fortuneteller's booth or parlor.

On the other hand, there is another side to crystal gazing. It has a very long and colorful history, and it has been practiced, in one form or another, in numerous cultures throughout the world. In this sense, crystal gazing does work, at least for certain people at certain times.

The equivalent of crystal gazing can sometimes be practiced using a bowl of clear water, and images have been seen in lakes and other bodies of water, polished stones, eyes, soap bubbles, and other shiny surfaces.

The most common form of crystal gazing, however, utilizes a smooth, flawless sphere of crystal, glass, or some precious transparent stone. Because of the high price of these materials, crystal balls made of acrylic are commonly sold, but these are considered unacceptable by authorities in the art of crystal gazing.

The finest crystal ball is still ineffective if the user does not have some degree of natural talent. Only a small percentage of the population possesses the ability to see images in the crystal ball.

If the user has this ability, certain techniques must

be followed to get optimum results. First, the surface of the sphere must be clean and free of any dust or fingerprints. If you are looking into the crystal for the benefit of someone that is with you, that person should be allowed to hold the crystal for a few minutes before you begin. The crystal should be held between the thumb and the other four fingers, or placed on a platform of appropriate size.

Before images appear in the crystal, it will become very hazy. Gradually, it will clear up, revealing the image. The image, however, seldom appears immediately. It may take a few seconds to as long as a half hour. Some crystal gazers place a piece of black cloth behind the crystal to minimize reflection and glare, whereas others prefer to concentrate on the reflections, claiming that this is where the images begin to appear.

The images that appear in a crystal ball may be things that are occurring in the present, did occur in the past, or will occur in the future.

In a famous case, Count Alessandro di Cagliostro (1743–1795) of Italy used the son of Chamberlain Count von der Howen for a public demonstration in crystal gazing. The boy gazed into a round glass carafe filled with water. He said that he could see his sister clutching her breast and then embracing her brother, who was a soldier. The Count did not understand the meaning of this vision until he returned home. He knew that his daughter was in good health, and was separated from his older son who was in active service.

Upon arriving at his home, the Count discovered that his daughter had experienced severe chest pains at the time of the vision, and that her brother was with her, having been granted a leave of absence.

This, of course, is an example of an event that took place at the time of the crystal gazing. The boy who saw the vision in the container was separated from his sister and his older brother by distance, but he was not separated by time.

In another case, reported to the *Journal of the Society for Psychical Research* (1923), Sir William Barrett describes an incident in which a crystal gazer, Miss Nell St. John Montague, saw a vision concerning a Mrs. Holt. Miss Montague looked into her crystal ball and saw Mrs. Holt's husband holding a revolver and talking on the telephone. He had a look of tragedy and despair, and kept looking at the door as if he expected someone to enter. The vision continues:

> He raises his right hand and staggers back, the revolver is now pointing at his own head—then I see blood everywhere gushing.

Miss Montague was horrified, and warned Mrs. Holt that a terrible tragedy would soon take the life of her husband. Mrs. Holt dismissed the whole matter as nonsense. Three days later, Mrs. Holt heard a shot, and found her husband dead, just as Miss Montague had predicted.

It would be possible to cite many other cases of accurate predictions by crystal gazers. The well-known psychic, Jeane Dixon, relies partially on crystal gazing in making her predictions. And many of her predictions have subsequently proved to be correct. She was able to foretell the election victories and death of Franklin Roosevelt, Dag Hammerskjold's plane crash, and the assassinations of John and Robert Kennedy and Martin Luther King.

The crystal ball apparently acts as one form of time machine which, under the right conditions, somehow enables one to view scenes and events from other time periods.

24 Magic Mirrors

Certain ancient writings on magic and sorcery alluded to magic mirrors through which one could transcend the barrier of time and perceive events in the past or the future.

Grillot de Givry, writing in his *Picture Museum of Sorcery, Magic, and Alchemy,* describes the "three vases of Artephius," a sort of time machine whose construction was outlined in an early French manuscript. According to the instructions given, three vases made of different kinds of earthenware are filled with oil of myrrh, wine, and water, respectively. Other details are given; the result is that the user should be able to see the past in one vase, the present in another, and the future in the third vase.

According to Andrew Tomas, a similar device was described by Franciscus Picus in the *Book of the Six Sciences.* He describes an instrument called the "Al Mucheff mirror," which is supposed to provide a panoramic view of the past, present, and future.

Another method, attributed to the great alchemist, Paracelsus (1493–1541), was described in detail by Franz Hartmann, M.D., in *Paracelsus and the Substance of His Teachings* (London, 1887):

> The electrum magicum [magic alloy] is prepared as follows: "Take ten parts of pure gold, ten of silver, five of copper, two of tin, two of lead, one part of powdered iron, and five of mercury. All these metals must be pure. Now wait for the hour when the planets

Saturn and Mercury come into conjunction, and have all your preparations ready for that occasion; have the fire, the crucible, the mercury and the lead ready, so that there will be no delay when the time of the conjunction arrives, for the work must be done during the moments of conjunction. As soon as this takes place, melt the lead and add the mercury, and let it cool. After this has been done, wait for a conjunction of Jupiter with Saturn and Mercury, melt the compound of lead and mercury in a crucible, and in another crucible the tin, and pour the two metals together at the moment of such conjunction. You must now wait until a conjunction of the sun with either one or both of the above-named planets takes place, and then add the gold to the compound after melting it previously. At a time of a conjunction of the moon with the sun, Saturn or Mercury, the silver is added likewise, and at a time of a conjunction of Venus with one of the above-named planets copper is added. Finally, at a time of such a conjunction with Mars, the whole is completed by the addition of the powdered iron. Stir the fluid mass with a dry rod of witch-hazel, and let it cool.

"Of this electrum magicum you may make a mirror in which *you may see the events of the past and the present* (author's italics), absent friends or enemies, and see what they are doing. You may see in it any object you may desire to see, and all the doings of men in daytime or at night. You may see in it anything that has ever been written down, said, or spoken in the past, and also see the person who said it, and the causes that made him say what he did, and you may see in it anything, however secret it may have been kept.

"Such mirrors are made of the electro [sic] magicum; they are made of the diameter of about two inches. They are to be founded at a time when a conjunction of Jupiter and Venus takes place, and moulds made of fine sand are used for that purpose. Grind the mirrors smooth with a grindstone, and polish them tripoly, and with a piece of wood from a linden tree. All the operations made with the mirror, the grinding, polishing, &c., should take place under favour-

> able planetary aspects, and by selecting the proper hours three different mirrors may be prepared. At a time of a conjunction of two good planets, when at the same time the sun or the moon stands on the 'house of the lord of the hour of your birth,' the three mirrors are to be laid together into pure well water, and left to remain there for an hour. They may then be removed from the water, enveloped in a linen cloth, and be preserved for use."

It is interesting to note that magic mirrors were not only used by the ancient alchemists, but are used by primitive tribes even today. Ernesto de Martino, writing in *The World of Magic,* quotes R. G. Trilles in *Les Pygmées de la forêt équatoriale:*

> "During one of our voyages . . . we arrived at a village one evening and met the witch-doctor who described to us in detail the route we had taken, the stops we had made, various meetings, the food we ate and even the conversation we exchanged. . . . We were speaking French, so the witch-doctor could not have understood a word, yet, without budging from his village, he had seen us in his magic mirror and repeated everything we said."

The above quotation was made in reference to a visit to a tribe of African Pygmies. Various other tribes throughout the world also employ magic mirrors, which are used by shamans or witch doctors as valuable aids to divination.

PART V: Theoretical Considerations

25 The Fourth Dimension

The difference between three spatial dimensions and four spatial dimensions is like the difference between a flat sheet of paper and a box. The sheet of paper has length and width but no depth. The box has length, width, and depth.

The universe in which we live is composed of three spatial dimensions, but let us suppose, for a moment, that we exist in a universe composed of two spatial dimensions. Everything would then have the appearance of sketches on a piece of paper. All things could be described in terms of length and width, but there would be no such thing as depth. Objects could move in any direction along the plane of the paper, but it would be impossible to move above or below the plane of the paper.

In this imaginary universe of two dimensions, suppose that a square was drawn around an object. Then that object could not escape from the square except through a hole or opening in the square. Movement over or under the square would be impossible.

Now let us return to our own universe of three spatial dimensions. If a square is drawn around a three-dimensional object, it is a very simple matter for that object to pass over or under the square. Now let us suppose that an object is enclosed in a cubic structure: say, a room with a ceiling, a floor, and four solid walls. That object cannot pass through the room unless there is an opening in the ceiling, the floor, or one of the four walls.

Now, this might all seem very obvious to anyone. Admittedly, it is possible to explain *almost* all phenomena in terms of a universe composed of three spatial dimensions. For instance, it is this principle which explains why water can be confined in a glass, or why a prisoner can be confined in a cell.

Let us now consider paranormal phenomena; for instance, materialization and dematerialization. One famous medium, Charles Bailey, was reputed to have materialized hundreds of objects into an iron cage in the presence of numerous skeptical witnesses. These objects were apparently made to pass through the bars of the iron cage which simply cannot be explained in terms of a three-dimensional universe. The strange case of the disappearance of David Lang, described elsewhere in this book, also seems to defy the three-dimensional concept.

As a possible explanation to paranormal phenomena such as that described above, it has been suggested that there exists a fourth spatial dimension that is normally, but not always, inaccessible. At times, objects are able to pass into and out of the fourth dimension.

One of the earliest research works which examined the concept of the fourth dimension was *Transcendental Physics* by Johann Carl Friedrich Zöllner. The original work was written in German, but an English version was published in Boston in 1881. For illustration, Zöllner cited the phenomena produced by Henry Slade, a medium. Slade was able to cause an object to disappear altogether, and then to reappear elsewhere. He was also able to cause two solid rings to materialize around the stand of a small table.

Slade was later imprisoned for fraud in England, which did irreparable damage to the reputation of Zöllner. In retrospect, however, it seems immaterial whether Slade was a fraud or not. Zöllner contributed a very carefully formulated theory to the world that has not lost its significance in the light of the scandal. And, it remains questionable whether or not Slade was actually a fraud.

The following quote is from *Transcendental Physics* by Zöllner:

> Among the proofs there is none so instructive and convincing as the transport of material bodies from a space enclosed on every side. Although for our three-dimensional intuition such a space apparently allows no other exit than through the material boundaries, yet from the fourth dimension it can be opened, and thus the transport of the body in *this* direction can be effected without disturbance of the three-dimensional material walls. Since the so-called intuition of a four-dimensional space is wanting to us, as three-dimensional beings, we can only form to ourselves a conception of this proceeding by an analogy taken from the next lower region of space. Suppose in a plane a figure of two dimensions enclosed by a line on every side, in which is a movable object. By movements *only* in the plane that object could not escape from the interior of that *two-dimensionally* enclosed space otherwise than by an opening of the line of enclosure.

In simple terms, Zöllner explains that an object can pass through solid matter only by means of four-dimensional movement. It is this movement, he says, that offers the most convincing evidence to date of the existence of a fourth dimension.

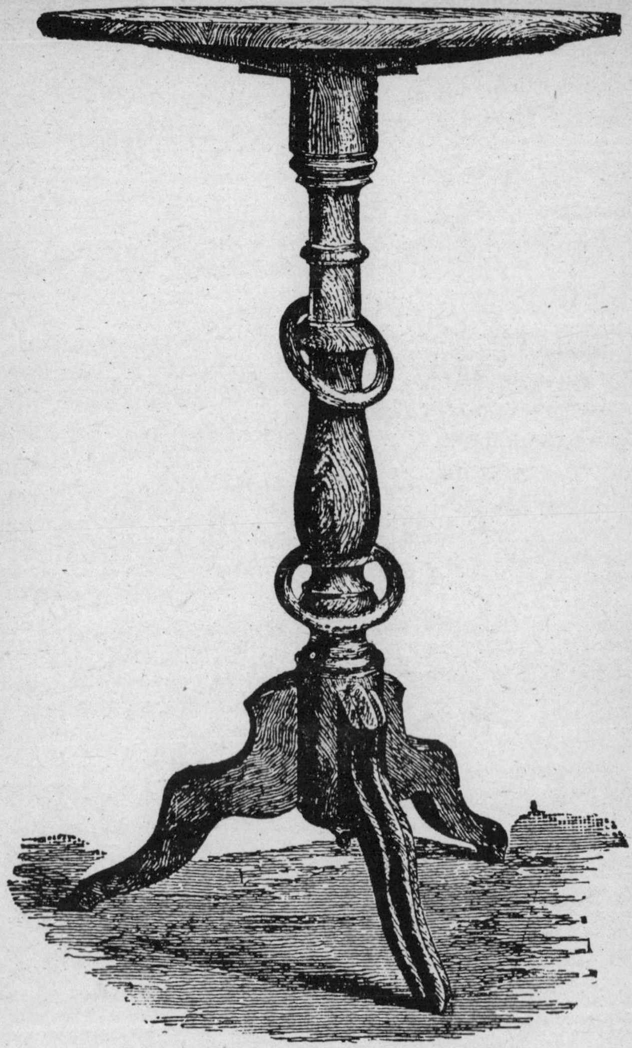

Illustration from *Transcendental Physics* by Zöllner depicting two rings that were materialized around the base of a small table by Henry Slade.

26 Time's Arrow

Imagine, for a moment, that you are watching a film of a game of pool. The game has just begun. The cue stick hits the cue ball; the cue ball breaks up the other balls. Some of the balls roll into the side pockets; others simply roll around and come to rest at various places on the pool table.

Now, imagine that the film is being run backwards. Several balls quickly leap out of the side pockets and roll toward the center of the table. The cue ball rolls back and comes to rest at the tip of the cue stick. All of the other balls come to rest, forming a neat triangle on the pool table.

Our experience tells us that the backward movements actually never occur. Though it may look amusing, we instinctively see this as an impossibility.

Suppose, however, that one was asked to explain why the backward movements are impossible? Which, if any, physical laws have been violated? At first, it would seem that gravity would have to be violated for the balls in the side pockets to roll back out. But suppose that the molecules that absorbed the shock and the heat of impact at the bottoms of the side pockets were to converge, and return the impulse to each of the balls, propelling them back up to the surface of the table.

These questions have concerned physicists for many years. In fact, the reversal of the entire course of events during a pool game would not violate any basic laws of physics, although the laws of probability would clearly be violated. The chances that such a thing would happen are very close to zero.

Until recent times, the laws of probability were considered to be fundamentally responsible for the fact that time reversal does not occur.

One of the most eloquent spokesmen for the probability explanation was Sir Arthur Eddington (1882-1944), a noted British astronomer and physicist. Writing in *The Nature of the Physical World,* he comments on the inevitable progression of time, which he calls "time's arrow":

> The great thing about time is that is goes on. But this is an aspect of it which the physicist sometimes seems inclined to neglect.

In describing a method of finding the direction of time's arrow, he stated:

> If as we follow the arrow we find more and more of the random element in the state of the world, then the arrow is pointing towards the future; if the random element decreases the arrow points towards the past.

On a smaller scale, this rule would apply quite appropriately to the example of the pool game. As the cue ball strikes the other balls, the neat, orderly triangle becomes randomly scattered. The random element is increased, and the arrow points toward the future. On the other hand, if the randomly scattered balls once again assume their positions in the triangle, the random element decreases, and the arrow points towards the past. In one case, time moves forward; in the other case, times moves in reverse.

On the basis of recent findings, it now appears that

there may be more fundamental reasons for the direction of time's arrow than probability.

On the atomic and subatomic levels, certain weak interactions between minute pieces of matter are apparently *not* time-reversible. In other words, it appears that these interactions always go in one direction, never the other way.

According to a related concept proposed by Richard P. Feynman, certain subatomic pieces of matter called antiparticles are actually particles traveling backward in time for a brief moment. In other words, what appears to be an antiproton is actually a proton traveling backward in time. What appears to be a positron is actually an electron traveling backward in time.

Feynman's concept is only a theory at this point; it has neither been proven nor disproven.

Nevertheless, it now appears that on a level beyond the power of the microscope, time's arrow *must* point in a forward direction in certain instances. If Richard Feynman is correct, time travel itself may be a common occurence at the submicroscopic level.

In any event, it would seem that scientists have just begun to unravel the fundamental principles that govern the direction of time.

27 Plato and Time Reversal

Writing in *The Statesman,** the Greek philosopher Plato mentions a peculiar phenomenon in which the universe is supposed to reverse its direction and move backward in time.

The description of time reversal given by Plato was offered as an explanation for a myth that was called to his attention. In the myth, the god Zeus was angered by an unjust king who stole the throne from his predecessor. By simply releasing his control over the universe, Zeus caused time to move in reverse, and in this way he restored the throne to the deposed king. Plato believed that the gods alternately control the universe, and release their control. Each cycle lasts for many centuries. While the gods control the universe, time moves forward. While the gods have released their control, time moves in reverse.

He describes the action of Zeus as follows:

> Listen then. There is an era in which the god [Zeus] himself assists the universe on its way and helps in its rotation. There is also an era in which he releases its control. He does this when its circuits have completed the due limit of the time thereto appointed. Thereupon it begins to revolve in the contrary sense under its own impulse—for it is a living creature and has been endowed with reason by him who framed it in the beginning.

**The Statesman* by Plato. Translated by J. B. Skemp; edited by Martin Ostwald. New York: Bobbs–Merrill Library of Liberal Arts. Copyright 1957 by The Liberal Arts Press. Reprinted by special permission.

Plato then describes the consequences of this strange time reversal:

> First of all, every living creature remained still at whatever stage of life it had attained. All mortal beings halted on their way to assuming the looks of old age, and each began to grow backwards, as it were, toward youth and even greater tenderness. The white hairs of the older men began to grow dark again; the cheeks of bearded men grew smooth and restored to each the long lost bloom of his youth. The bodies of the young men lost the signs of manhood and, growing smaller every day and every night, they returned again to the condition of newborn children, becoming like them in mind as well as in body. Next they withered away completely and were wiped out altogether.

The idea of a universe moving backward in time raises some interesting questions.

For instance, in the preceding example, the mythical universe first went forward in time, then reversed its direction and went backward in time. Presumably, the inhabitants of that universe were not aware that time was moving in reverse, although Zeus was well aware of it.

At Mount Olympus (the legendary home of Zeus and the other Greek gods), it would follow that time continued to proceed in a forward direction. Otherwise, Zeus and the other gods would have also moved back in time.

Although the example is mythical, it raises an important question. Does time simply move backward or forward, or is the direction of time relative to the observer? Can time move forward from one point of view and backward from another?

Long after the time of Plato, this question was considered by Albert Einstein. The conclusions he arrived at are discussed in the next chapter.

28 The Theory of Relativity

The theory of relativity, a milestone in man's understanding of the universe, resulted largely from the work of the late Albert Einstein. Consistent with the theory is the conclusion that time travel is a very real possibility for objects traveling at tremendous speeds.

According to the theory of relativity, no object moves at an absolute speed; the speed of an object is relative to the observer. For example, suppose that a space ship were launched toward the moon. Relative to the earth, the space ship might be receding at, say, 1000 miles an hour, and approaching the moon at 1000 miles an hour. Relative to the space ship, it would be equally true to say that the earth is receding at 1000 miles an hour, while the moon is approaching at 1000 miles an hour. Which observation is correct? According to the theory, both observations are equally valid from their respective points of reference.

Another basic premise of Einstein's theory involves the speed of light. Light, according to Einstein, travels at an absolute speed that is *not* relative to the observer. Suppose that a strong beam of light is directed into outer space. The light will travel away from the earth at 186,000 miles a second. At the same instant, suppose that a space ship is launched on a parallel course with the beam of light, but traveling slightly under the speed of light, say, 176,000 miles a second. Relative to the space ship, would the beam of light appear to be traveling ahead at 10,000 miles a second? The answer is no. The light beam would always appear to be traveling at the speed of 186,000 miles per second, regardless of the speed or position of the observer.

So far, we have two basic premises: (a) the speed of any object is relative to the observer, and (b) the speed of light is not relative to the observer; it is always constant.

For both of the above to be correct, the space ship that is traveling at 176,000 miles per second would undergo strange transformations. As the speed of light is approached, the mass* of the space ship and everything inside it would increase. Due to the increased mass, all moving parts would move slower. Clocks aboard the space ship would run slower so that the beam of light would always appear to have a speed of 186,000 miles a second. If a measuring stick were somehow used to determine the distance traveled by the beam of light during a certain length of time, the measuring stick would actually become shorter; consequently, the beam of light would appear to have traveled farther.

To an observer on earth, the clocks aboard the space ship would seem to be running slow. To a person aboard the space ship, the clocks in the ship would appear to be normal, but the clocks on earth would seem to be running fast. From earth, the measuring stick aboard the space ship would appear too short. From the space vessel, the measuring sticks on earth would seem to be too long.

This brings us to the effect of relativity on the passage of time. From the earth, people aboard the space ship would seem to be aging very slowly, and time would seem to pass very slowly. From the space ship, time would seem to be moving very rapidly on earth. The people aboard the space vessel might be away for,

*mass: the amount of matter present in an object

say, ten years, only to return and find that the equivalent of 100 years had passed on earth.

According to the theory of relativity, an object could never attain the speed of light because its mass would become infinite while its length would shrink to nothing. This is not possible within the limits of our present knowledge.

From a practical standpoint, no space ship has ever even approached the speed of light. Thus, it is not possible to actually test these theoretical conclusions. Nevertheless, it is certain that some form of time travel would occur if these speeds could be attained.

The problem of straight-line motion at any speed is that the object would not remain in the same place. For instance, if a space ship were launched upward from the earth at the speed of light, it would be extremely difficult, if not impossible, to measure differences in the passage of time on earth and the passage of time aboard the space ship.

Theoretically, however, an object need not travel in straight-line motion at tremendous speeds for time differences to be noticeable. For instance, if a sphere could be induced to rotate or vibrate at great speeds, the same principles of relativity would apply. A few hours might go by inside the sphere, while several years might pass outside the sphere.

To build such a structure that could withstand either vibrational or rotational motion at speeds near the velocity of light would be an exceedingly difficult task. The sphere would have to be perfectly constructed so that it was perfectly round and perfectly balanced. The slightest flaw—the tiniest nick—would cause the sphere to break apart under the strain long before the desired speed was attained. To illustrate this point, consider that the tires on your car are balanced for speeds up to, say, seventy miles an hour. At greater speeds, your car would start to shake. A sphere rotating at the speed of light would be moving at 111,600,000 miles an hour. Our present technology is not capable of balancing anything with such precision.

Even if the balancing were possible, the sphere would have to be protected against friction, the bombardment of tiny dust particles, and internal strain. Friction could be minimized by coating the sphere with a frictionless substance such as Teflon and placing it in a vacuum. To eliminate the danger of bombardment by tiny particles, a force field capable of repelling matter would have to be created around the sphere, something that is still beyond present technology. To eliminate the problem of internal strain, the sphere would have to be made out of an extremely resilient substance. It is not known whether any substances known to man would be durable enough; the strongest alloys have never been subjected to this kind of treatment.

Finally, the greatest problem would be to make such a contraption safe enough for a living passenger. I will not even attempt to speculate on how this might be done.

Difficult as the task may seem, it should not be considered an impossibility. A century ago, the laser and the electron microscope would have been considered well beyond the realm of possibility; today they are a reality.

29 A Look into the Past

Have you ever gazed up at the sky on a very clear, dark night when millions of stars were visible? If so, then you have also peered back through time into the distant past.

Other than the sun, the star that is nearest to the earth is Proxima Centauri, which is about 24 trillion miles away. Light, which travels at 186,000 miles per second, takes four years to reach the earth from Proxima Centauri. Thus, Proxima Centauri is said to be four light-years away. What *appears* to be Proxima Centauri is actually Proxima Centauri as it was four years ago. Suppose that this star were destroyed today in a great cataclysmic explosion. Then the explosion would not be visible on earth for another four years.

Only a handful of stars are within fifteen light-years of our solar system; some visible stars are billions of light-years away. Depending on which star one happens to be observing at a particular moment, a viewer may be watching a star as it was four years ago, or as it was billions of years ago.

Let us suppose that someone on a very distant planet possessed some means of viewing the earth from thousands of light-years away. This situation was conjectured by Richard A. Proctor, writing in *Other Worlds Than Ours* (1870):

> To a being placed on some far-distant orb, whence light would occupy thousands of years to wing its flight to us, there would be presented, if he turned his

> gaze upon our earth, and if his vision were capable of telling him of her aspect, the picture of events which thousands of years since really occurred upon her surface. For the light which left the earth at that time, winging its way through space with the account, if we may so speak, of those occurrences, is now travelling as swiftly as when it left our earth, but amid regions of space removed from us by a light-journey thousands of years in duration. And thus, to the observer on this distant orb, the events which happened in the far-off years would seem to be actually in progress.

Today, most astronomers feel that light from any planet would be so greatly scattered and diffused after traveling such a great distance in space, that not even the most perfect telescope would permit a view of that planet.

Admittedly, this accomplishment is still beyond present-day technology. However, it is interesting to note that this achievement was predicted in *A Dweller on Two Planets,* which is discussed elsewhere in this book. The following quotation appeared in a later version (1899) of *A Dweller on Two Planets* in the author's preface:

> He [Phylos] said that electric currents unimpressed with vibrations such as [those that] produce sound, heat and light, until resisted, would be superadded to the light vibrations constituting the image beheld through the telescope. . . . Thus he stated that upon suns and heavenly bodies so distant that hundreds of them only (even in this year A.D. 1899) seem as a faint speck through the most powerful telescopes, to this electrostellarscope would, by proper amplification of the electro-luminous waves, be made so plain to earthly vision that objects not visible to unaided earthly sight would be easily perceptible on the most distant stellar body, however remote from the mundane beholder.

Apparently, this refers to a yet unknown method of enhancing the telescopic image through the use of electric current. To date, the only significant method to improve telescopes is to utilize larger and more perfect lenses in their construction.

Perhaps future developments in the field of astronomy will afford mankind a more penerating view into the vast time tunnel that is above us.

Afterword

Imagine how eons ago primitive man watched with awe and terror as a bolt of lightning struck a nearby tree and set it afire. Having no knowledge of electricity, he could not comprehend the lightning. Without a knowledge of the nature of fire, he could not comprehend the burning tree. Similarly, modern man is mystified to hear of accounts of people who vanished suddenly and mysteriously in the presence of others and never returned. Yet those who did return—such as Mrs. W. of Tunbridge Wells, Miriam Golding, Joseph Sainsbury, the Petersons, and J.P.J. Chapman—might have given us a few clues which point to the existence of other dimensions that are yet uncharted. These experiences are not too common, and the less adventurous souls among us would just as soon prefer to be spared a sudden encounter with an unknown dimension.

Far more common, however, are dreams, premonitions, and related phenomena in which encounters with other dimensions or time periods are limited and temporary. Sometimes these dreams and premonitions are so intense and full of detail that they are more properly termed visions. George Washington experienced a vision at Valley Forge which seemed to foretell the future of what was then an infant nation. Two centuries before, Nostradamus had envisioned the future of the world as stated in his *Centuries*. And long before that St. John was given a vision of the future which has been handed down in the obscure language of the *Book of Revelation* (*Apocalypse*).

Similarly, certain writers and scientists have envisioned the future with startling detail and accuracy that would indicate more than just an active imagination.

But these experiences with other dimensions and time periods are, for the most part, entirely unsolicited. They seem to just happen, though they may be subtle or startling, insignificant or profoundly important. One might naturally wonder if time travel, like the weather, will forever elude man's conscious control. Will a time machine like that envisioned by H. G. Wells ever be constructed, or must we be content to take trips across the time barrier whenever fate determines?

If indeed man is destined to someday build a workable time machine, why has there not been more tangible evidence of such vehicles landing in the past? A few unexplained fossil footprints and historical enigmas are all that we can point to as evidence that future man has walked the earth. Could it be that future man will be wisely cautious and use the time machine with great discrimination? Perhaps, then, the fossil footprints were mistakes—traces that future man left in spite of his meticulous efforts to leave none.

It is a fact that present-day researchers who have taken a serious look at the possibility of constructing a machine have two major disadvantages working against them—if they lack credibility, they are apt to be ridiculed and scorned; but if they seem to be on the trail of a genuine discovery, it then becomes a matter of national security. It is highly unlikely that the government would ever permit private individuals to possess workable time machines any more than the government would allow the public to buy and sell atomic bombs. In fact, the United States government has a complete monopoly on all radioactive elements and inert gases within its boundaries. They can be bought and sold only in accordance with government regulations. There is little doubt that the government would act immediately to regulate the use and possession of time machines.

I know of one man, whose name I will not mention, who claims to have built a time machine which he offered for sale to the public several years ago. He claimed that the unit would send the user into the past or future, but he cautioned that the only drawback was that one could not determine which period in history he would visit. I did not personally see the machine in operation, so I cannot comment on its efficacy.

To the surprise of no one, this man was put out of business by the Postal Service, which labeled the device a fraud. In all fairness, however, it should be noted that the Postal Service does not normally bother to investigate such claims, but places the burden of proof entirely on the claimant. The claimant is, in effect, guilty of fraud unless he can satisfactorily prove himself innocent.

To cite another example, George Van Tassel does not offer a time machine for sale to the public, but he does admit that the still unfinished Integratron will be, among other things, a time machine. He, too, has been subject to government scrutiny. He states that he has been visited by the Defense Department. Bryant and Helen Reeve, after visiting Van Tassel said, "We noted that one of the names on the mailing list of *Proceedings* was Uncle Sam's F.B.I.!"

George de la Warr perhaps did not fall victim to a significant amount of government harassment, but he ran into difficulties in gaining acceptance among doctors and physicists that would have halted the work of all but the most determined of pioneering souls. Some physicians actually obtained impressive results using de la Warr's methods but abandoned their association with de la Warr simply because they feared reprisal from the medical profession.

Leonardo da Vinci was too far ahead of his time to be appreciated by his contemporaries. He was truly one of the most farseeing men of history. Though he lived and worked in the fifteenth and sixteenth centuries, his mind seemed to dwell in the twentieth century and beyond. His greatness is only appreciated in retrospect by people of more modern times.

Nikola Tesla was also far ahead of his time and was neither appreciated by his contemporaries, nor is his greatness truly appreciated today. History has erroneously attributed many of his accomplishments to the work of Thomas Edison. Though Edison was a great man in his own right, he merely popularized some of the discoveries actually made by Tesla. A comparison of Edison and Tesla gives us a clue that might explain Edison's greater appeal and acceptance. Thomas Edison was an all-American lad who showed early signs of genius, and grew up to become a respected scientist and inventor. Nikola Tesla, however, was born in what is now Yugoslavia and came to live in New York City. (The tourist who visits Eastern Europe might wish to include the Nikola Tesla Museum in Beograd, Yugoslavia, in his itinerary.) Tesla, though a kind and gentle man, was not without certain eccentricities. It is said that he would have nothing to do with women on a social basis, and he shied from the advances of even the famous French actress Sarah Bernhardt. He loved pigeons, and would bring home those that looked sickly to nurse back to health in his desk drawer. He was definitely a loner, and was unable to work closely with others.

Tesla was definitely an individual. By some standards his lifestyle would be considered weird. Yet even those who can accept Tesla and his achievements have trouble imagining a force field that could render a nation impermeable to attack, or a workable antigravity machine.

Having now looked at isolated instances of encounters with other time periods and dimensions, and having looked at writers, scientists, and inventors who either saw into the future, or involved themselves in some aspect of time travel research, we next turn to philosophy and theoretical physics for a possible answer. Is time travel a theoretical possibility? Are there known laws of the universe that say it can or cannot be done? In perusing the works of Plato, Einstein, and others, we obtain an answer in the affirmative. Time travel does not violate any known laws of the universe; in fact, philosophy and the theory of relativity offer possible clues as to how time travel might be explained.

The bulk of evidence would indicate that time travel is not a myth or an impossibility, but rather a pioneer field with the potential to bring man to horizons yet undreamed of. In this endeavor, both the philosopher and the scientist must combine their resources and efforts in the search for a true understanding of time travel, and a greater level of awareness for all mankind.

Glossary

Certain words have been used in this book which may be unfamiliar to some readers. The following definitions have been added to clarify the meanings of these words. Terms which have been defined elsewhere in the book are not listed below.

clairaudience An ability to hear sounds that are normally imperceptible to human hearing. These sounds may orginate from other dimensions, other time periods, or other planes of existence. Clairaudience is the perception of sounds that originate outside the subject and should not be confused with auditory hallucinations which originate within the subject and manifest an abnormal mental state.

clairvoyance An ability to see things that are normally imperceptible to human sight. The persons, events, or objects that are seen originate from other time periods, or other places.

clairvoyance, contemporaneous An ability to see persons, events, or objects that are located at a remote place, but *not* in a different time period. Events are witnessed at the time they occur; persons and objects are perceived just as they are at the time of the experience.

clairvoyance, precognitive An ability to see things from a future time period.

clairvoyance, retrospective An ability to see things from the past.

Fortean fall A term derived from the name of Charles Fort (1874-1932), who devoted his life to the investigation of paranormal phenomena.

A Fortean fall refers to the fall of any abnormal substance from the sky that cannot be explained by science. Nearly every imaginable substance has, in fact, been reported at one time or another, including alligators, live frogs, fish, sacks of cookies, red meat, and blood. Perhaps these objects fall through some sort of interdimensional warp in time and space and enter into our dimension. However, this is only a theory—one of very few theories that have been offered as a possible explanation.

hallucination the apparent perception of sights, sounds, smells, tastes, and tactile sensations that are not actually present. Hallucinations originate in the mind of the individual who experiences the hallucination. They are brought on by psychosis, the ingestion of an hallucinogen, or various abnormal physical states caused by high fever, prolonged fasting, extreme pain, and other factors.

Hallucinations may be visual (sight), auditory (sound), olfactory (smell), gustatory (taste), or tactile (touch). They may be very concrete and lifelike, or very vague and ethereal. They are not considered to be in the realm of psychic or paranormal phenomena, and can usually be explained by science.

hallucinogen Any substance capable of inducing hallucinations.

There are natural hallucinogens which occur in nature. Most of these are present in certain types of plants, although some animal venoms are also hallucinogenic. Plant hallucinogens include mescaline, from peyote *(Lophophora williamsii),* and psilocybin, from the mushroom *Psilocybe mexicana.* For reasons that are not yet known, the overwhelming majority of plant hallucinogens are native to Central and South America.

There are also numerous synthetic hallucinogens, such as LSD, MDA, and others, that are man-made and do not occur in nature.

PK effect See: psychokinetic effect.

precognition The perception of an event before it actually occurs.

premonition An experience of something prior to its actual occurence. To experience a premonition is a form of precognition.

psychokinesis A power which enables a person to move something by the exertion of mental force.

psychokinetic effect (PK effect) The movement of an object resulting from psychokinesis. Examples of psychokinetic effects include levitation (raising an object from the ground) and the movement of objects across a table by intense concentration.

psi phenomenon See: psychic phenomenon.

psychic (psi) phenomenon Any type of extraordinary phenomenon that cannot be explained, as yet, by science.

retrocognition The perception of an event after it has already occurred; the perception of an event that occurred in the past.

retrospection Same as retrocognition.

telepathy Mental communication. Telepathy refers to the ability to communicate with another individual, often over great distances, without the use of any mechanical device.

Telepathy may involve receiving thoughts from another individual, sending thoughts to another individual, or both. Telepathy is probably the fastest form of communication known. There is apparently no delay whatsoever in receiving a thought after it has been sent.

Bibliography

Allen, W. Gordon. 1959. *Space-Craft From Beyond Three Dimensions*. New York: Exposition Press.

Baudelaire, Charles. 1956. "The Seraphic Theatre." In *The Essence of Laughter and Other Essays,* edited by Peter Quennell and translated by Norman Cannon. London: Weidenfeld and Nicolson.

Bellamy, Edward. 1951. *Looking Back*. New York: Random House (Modern Library).

Bierce, Ambrose. 1893. *Can Such Things Be?* New York: Caffell Publishing Co.

Blatt, John M. 1956. "Time Reversal." *Scientific American* 195:107–114.

Bragdon, Claude. 1923. *Four-Dimensional Vistas*. New York: Alfred A. Knopf.

Burges, George A. 1881. *The Works of Plato*. London: George Bell & Sons.

Clarke, Arthur C. 1972. *Profiles of the Future*. New York: Bantam Books.

Corelli, Marie. 1921. *The Secret Power*. London: Methuen & Co.

Crabb, Riley Hansard. 1974. "Daisy, the Tube (Part I)," *Journal of Borderland Research* 30, No. 2 (March–April 1974):9–14.

———. 1970. "The White Light Technique for Youth," *Journal of Borderland Research* 26, No. 1 (January–February 1970):22–23.

———. 1970. "Who Wants to Be a Chrononaut?," *Journal of Borderland Research* 26, No. 1 (January–February 1970):30–31.

Denton, William and Elizabeth. 1866. *The Soul of Things*. Boston: Walker, Wise, & Co.

Eddington, Sir Arthur. 1958. *The Nature of the Physical World.* University of Michigan Press: Ann Arbor Paperbacks.

Edwards, Frank. 1964. *Strange World.* New York: Bantam Books.

———. 1959. *Stranger Than Science.* New York: Bantam Books.

Ellwood, Garcia Fay. 1971. *Psychic Visits to the Past.* New York: New American Library (Signet Mystic).

Fenyo, Eva. 1959. *A Guided Tour Through Space and Time.* Englewood Cliffs, New Jersey: Prentice-Hall.

Fernandez, James W. 1972. "*Tabernanthe iboga*: Narcotic Ecstasis and the Work of the Ancestors." *Flesh of the Gods.* New York: Praeger.

Gardner, Martin. 1967. "Can Time Go Backward?" *Scientific American* 216:98–108.

Goudsmit, Samuel A.; Robert Clairborne; and the editors of *Life.* 1966. *Time.* New York: Time, Inc.

Greenhouse, Herbert B. 1971. *Premonitions: a Leap into the Future.* New York: Warner Books.

Hinton, C. Howard. 1904. *The Fourth Dimension.* London: Swan Sonnenschein & Co., Ltd.

Holzer, Hans. 1969. *Window to the Past.* Garden City, N.Y.: Doubleday.

Huxley, Aldous. 1968. *The Doors of Perception and Heaven and Hell.* London: Chatto & Windus.

Kaufman, Gerald Lynton. 1938. *The Book of Time.* New York: Julian Messner, Inc.

Keel, John A. 1971. *Our Haunted Planet.* Greenwich, Conn.: Fawcett Publications.

Lethbridge, T. C. 1967. *A Step in the Dark.* London: Routledge and Kegan Paul.

Marder, L. 1971. *Time and the Space Traveller.* London: George Allen & Unwin Ltd.

Meerloo, Joost A. M. 1970. *Along the Fourth Dimension.* New York: John Day Co.

Murphy, Gardner. 1967. "Direct Contacts with Past and Future: Retrocognition and Precognition." *Journal of the American Society for Psychical Research* 61:3–19.

O'Neill, John J. 1971. *Prodigal Genius.* New York: David McKay (Tartan Books).

Ponciano, Joe, and Ray Hudd. 1965. "Project Hermes Revived." *Journal of Borderland Research* 21, No. 5 (July–August 1965):5–12.

Russell, Bertrand. 1956. *The ABC of Relativity.* New York: New American Library (Mentor Books).

Sanderson, Ivan T. 1970. *Invisible Residents.* New York: Avon Books.

Spencer, John Wallace. 1973. *Limbo of the Lost.* New York: Bantam Books.

Steiger, Brad. 1974. *Mysteries of Time and Space.* Englewood Cliffs, N.J.: Prentice-Hall.

———. 1972. *Strange Disappearances.* New York: Lancer Books.

———, and Loring G. Williams. 1971. *Minds Through Space and Time.* New York: Award Books.

Steinberg, Saul. 1974. "Improving the Nick of Time." *Harper's* 248, No. 1484 (January 1974):3–10.

Storm, Margaret. 1959. *Return of the Dove.* Baltimore: Margaret Storm.

Suppes, Patrick. 1973. *Space, Time, and Geometry.* Dordrecht, Holland: D. Reidel Publishing Co.

Tomas, Andrew. 1971. *We Are Not the First.* New York: Bantam.

Tuttle, Hudson. 1900. *Mediumship and Its Laws.* Chicago: Regan Printing House.

Vallentin, Antonina. 1938. *Leonardo da Vinci: the Tragic Pursuit of Perfection.* Translated by E. W. Dickes. New York: Viking Press.

Van Tassel, George W. 1958. *Religion and Science Merged.* Yucca Valley, Calif.: Ministry of Universal Wisdom.

Wells, H. G. 1895. *The Time Machine.* London: William Heinemann.

Whitrow, G. J. 1972. *The Nature of Time.* New York: Holt, Rinehart, and Winston.

Wilkins, Harold T. 1958. *Strange Mysteries of Time and Space.* New York: Ace Books.

Zöllner, Johann Carl Friedrich. 1881. *Transcendental Physics.* Translated by Charles Carleton Massey. Boston: Colby & Rich.